The Minerals, Metals & Materials Series

Nihad Dukhan
Editor

Proceedings of the 11th International Conference on Porous Metals and Metallic Foams (MetFoam 2019)

TMS

Springer

Editor
Nihad Dukhan
Department of Mechanical Engineering
University of Detroit Mercy
Detroit, MI, USA

ISSN 2367-1181 ISSN 2367-1696 (electronic)
The Minerals, Metals & Materials Series
ISBN 978-3-030-42797-9 ISBN 978-3-030-42798-6 (eBook)
https://doi.org/10.1007/978-3-030-42798-6

This Springer imprint is published by the registered company Springer Nature Switzerland AG
The registered company address is: Gewerbestrasse 11, 6330 Cham, Switzerland

Preface

MetFoam 2019 is the 11th edition of the biannual International Conference on Porous Metals and Metallic Foams; it was heled in Dearborn, Michigan, USA between August 20 and 23, 2019. Previous editions were held in various cities around the world starting in Bremen, Germany in 1999. This conference is the largest cross-disciplinary, international technical meeting focused exclusively on the production, properties, and applications of these lightweight multifunctional porous materials. In Dearborn, a total of 115 presentations were delivered at the conference by participants from 17 countries. Many participants travelled long distances to present and attend the conference. Students delivered 40% of the presentations. Emerging areas such as additive manufacturing and freeze casting, as well as nanoporous materials, cellular materials, and metallic metamaterials were covered by some of the presenters.

The proceedings volume of MetFoam 2019 is the culmination of several months of work, which included the preparation of the papers by the authors, editing, and reviewing. The papers collected in this volume provide a representative snapshot of the research activity in the field. It is my sincere hope that this proceedings volume will remain a valuable record of MetFoam 2019, and that it will serve as a reference for researchers interested in porous metals and metallic foams.

I would like to thank the reviewers and the session chairs for volunteering their precious time and effort. The International Scientific Board and the Steering Committee of MetFoam 2019 provided indispensable input and guidance throughout the planning of the conference. Thanks are also due to Kelcy Wagner and Trudi Dunlap from our organization partner The Minerals, Metals & Materials Society (TMS) for their assistance in the production of the proceedings volume, as well as for their vital hard work during the organization phase of the conference. Certainly, a special thanks go to our sponsors and exhibitors for making the event possible.

Nihad Dukhan
General Chair

Contents

Part I
Design of Porous Metals

Design of Energy-Absorbing Materials for Space Crafts Based on Voronoi Diagrams

Koichi Kitazono, Keiji Matsuo, Takuya Hamaguchi and Yuta Fujimori

Abstract Energy-absorbing parts are important for the landing system of space crafts on celestial bodies. Open-cell porous aluminum alloys for energy as an energy-absorbing material were manufactured through the powder bed fusion (PBF) process. Mechanical properties of the porous metals depend on the cell structure as well as the porosity. The authors manufactured ordered cell structures, truncated octahedron and rhombic dodecahedron, and disordered cell structure, based on Voronoi diagrams using a 3D-CAD software. Compression tests revealed the high compressive strength and low energy absorption efficiency of porous aluminum alloys with ordered cell structure. On the other hand, the porous aluminum alloys with disordered cell structure showed relatively high energy absorption efficiency. Controlling the disordered cell structure designed by a 3D-Voronoi diagram enables to develop the advanced porous metals having various mechanical properties.

Keywords Additive manufacturing · Voronoi diagram · Porous aluminum · Energy absorption

Introduction

Porous metals are lightweight cellular materials having closed- or open-cell structures [1]. Aluminum foams are focused on as an energy-absorbing material for transport industries [2]. The purpose of the present study is to develop an energy-absorbing landing part of space crafts. The authors developed open-cell porous aluminum alloys manufactured through additive manufacturing (AM) process [3]. The AM technology is typically focused on titanium alloys due to their high specific strength and biocompatibility [4]. Recently, aluminum alloy parts can be manufactured through a powder bed fusion (PBF) method [5]. Most previous researches on the AM aluminum lattice structure have been considered ordered and periodic cell structures [6]. On the other hand, classical aluminum foams manufactured through cast or powder

K. Kitazono (✉) · K. Matsuo · T. Hamaguchi · Y. Fujimori
Tokyo Metropolitan University, 6-6 Asahigaoka, Hino, Tokyo 191-0065, Japan
e-mail: kitazono@tmu.ac.jp

N. Dukhan (ed.), *Proceedings of the 11th International Conference on Porous Metals and Metallic Foams (MetFoam 2019)*, The Minerals, Metals & Materials Series, https://doi.org/10.1007/978-3-030-42798-6_1

metallurgical processes have disordered and heterogeneous cell structures [7]. The authors use a method of 3D-Voronoi diagram [8] in order to design both ordered and disordered cell structures.

If the seed points are arranged to the random distribution, the resultant cell structure becomes disordered. If the seed points are arranged to the periodic and regular distribution, the resultant cell structure becomes ordered. The present study focusses on the difference of compressive properties between AM porous aluminum alloy specimens having ordered and disordered cell structures.

Design of Cell Structure

Two types of ordered cell structures, bcc-Voronoi and fcc-Voronoi, are considered in this study (Fig. 1). Arrangements of seed points of 3D-Voronoi diagrams are bcc and fcc lattice points, respectively. Their unit-cell structures are truncated octahedron with 36 struts and rhombic dodecahedron with 24 struts, respectively.

The nominal porosity of each cell structure can be calculated geometrically. It is assumed that the shape of a strut is a cylinder with the diameter of t and the length of a, and both ends of a strut are circular cones with the height of $t/2$. Volumes of the unit cells are expressed as

$$V_{\mathrm{bcc}}^{0} = 8\sqrt{2}a^{3}, \tag{1}$$

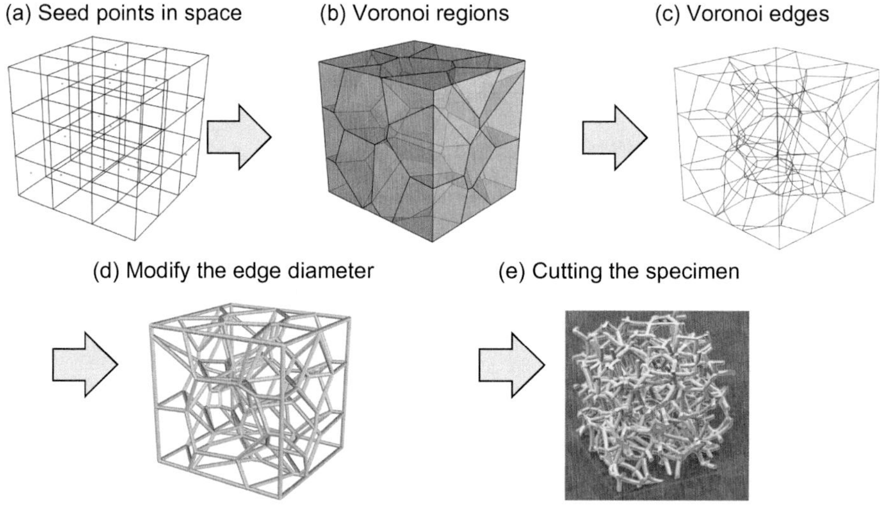

Fig. 1 Construction process of 3D-Voronoi diagram and open-cell structure

$$V_{\mathrm{fcc}}^{0} = \frac{16}{9}\sqrt{3}a^{3}. \tag{2}$$

Then, the nominal porosities of bcc- and fcc-Voronoi cell structures are calculated as

$$p_{\mathrm{bcc}} = 1 - \frac{3\sqrt{2}\pi}{16}\left(\frac{a}{t}\right)^{-2}\left\{1 - \frac{2}{3}\left(\frac{a}{t}\right)^{-1}\right\}, \tag{3}$$

$$p_{\mathrm{fcc}} = 1 - \frac{3\sqrt{3}\pi}{8}\left(\frac{a}{t}\right)^{-2}\left\{1 - \frac{2}{3}\left(\frac{a}{t}\right)^{-1}\right\}. \tag{4}$$

These equations are identical to the Ashby's equation [9] for open-cell foams expressed as

$$p = 1 - C\left(\frac{a}{t}\right)^{-2}, \tag{5}$$

where C is the constant.

In Fig. 2, nominal porosities of bcc- and fcc-Voronoi cell structures are plotted as a function of $a/t > 1$. Nominal porosities increase with increasing the strut length. At the same nominal porosity and strut thickness, the length of the strut in the fcc-Voronoi cell becomes longer than that in the bcc-Voronoi cell. Therefore, the cell size in the fcc-Voronoi cell structure becomes larger than that in the bcc-Voronoi cell structure. Using these relationships, ordered porous metals having optional porosities can be designed.

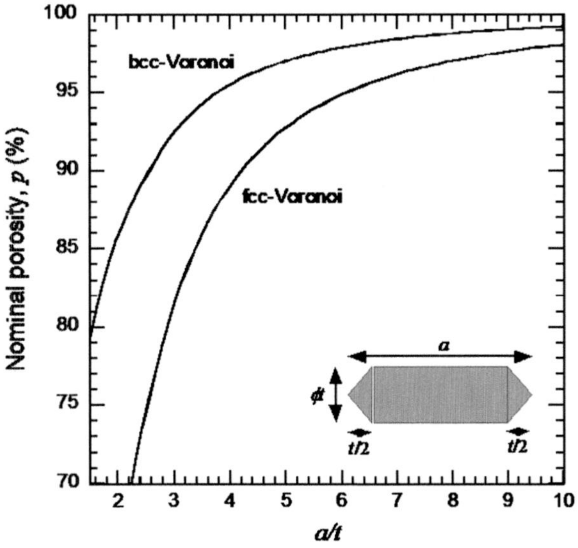

Fig. 2 Nominal porosities of bcc- and fcc-Voronoi cell structures are plotted as a function of *a*/*t*

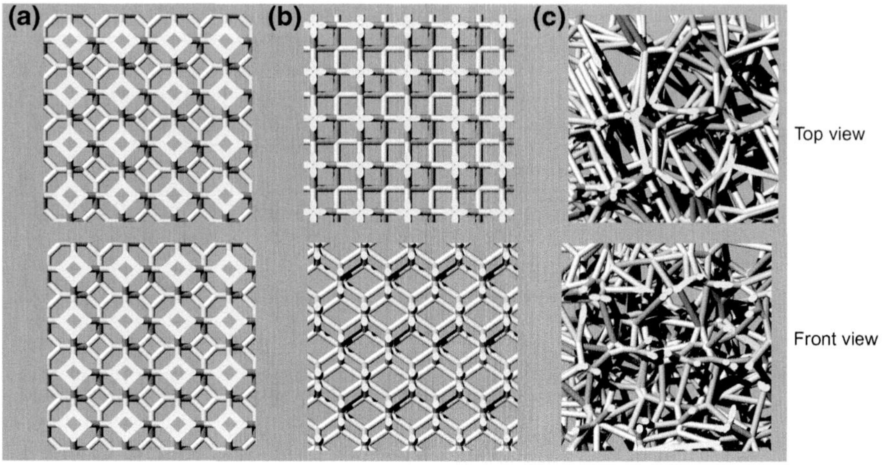

Fig. 3 3D-CAD images of **a** bcc-Voronoi [001] cell, **b** fcc-Voronoi [001] cell, and **c** random Voronoi cell

All 3D-CAD data were constructed by a 3D-CAD software, Rhinoceros 6. For disordered cell structure, the random cell structure was designed by random arrangements of seed points. Three 3D-CAD images of cubic specimens are shown in Fig. 3. Nominal porosities are 90% in bcc-Voronoi and fcc-Voronoi cell structures and 93% in random cell structures, which are calculated from the volume of the solid elements. All cell struts have cylindrical shape with 1 mm in diameter. Since the bcc-Voronoi cell has fourfold rotation symmetry, the top view is the same as the front view.

Experimental Procedure

Compression specimens were manufactured through a PBF machine, EOS M280, in Koiwai Co., Ltd., Japan. Al–10Si–0.3 Mg alloy powder is used as a starting material. Building direction is parallel to the z-axis. Specimens have cubic shape with a side length of 30 mm (Fig. 4). As-built specimens were annealed at 803 K for 6 h in air. Bulk density, ρ^*, was calculated from the total mass and the volume of the cubic. The porosity of each specimen is calculated as

$$p = 1 - \frac{\rho^*}{\rho_S}, \qquad (6)$$

where ρ_S is the density of dense Al–10Si–0.3 Mg alloy, 2670 kg/m^3. Because of the accuracy of the present PBF process, thickness of the cell struts in AM porous aluminum alloys was smaller than that of 3D-CAD models. As a result, the porosity of AM specimens became slightly high.

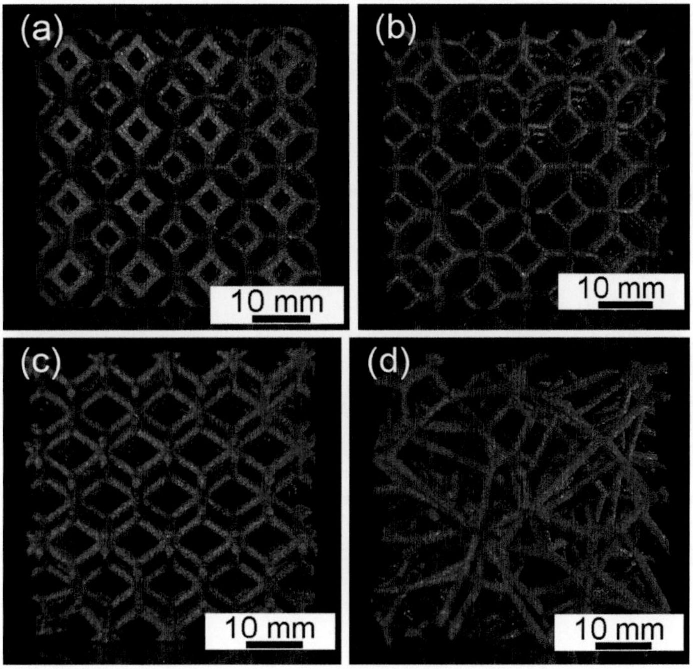

Fig. 4 Images of porous Al–10Si–0.3 Mg alloy specimens, **a** bcc-Voronoi [001] cell, $p = 91\%$, **b** bcc-Voronoi [001] cell, $p = 94\%$, **c** fcc-Voronoi [001] cell, $p = 91\%$, **d** random Voronoi cell, $p = 94\%$

Compression tests were carried out using a Shimadzu Autograph AG-50kNISD universal testing machine. Crosshead speed was fixed at 10 mm/min. No lubricant was used in compression tests.

Results and Discussion

Results of compression tests are shown in Fig. 5. Compression direction of the bcc-Voronoi cell [001] is parallel to the building direction. Flow stress increased with decreasing the porosity. In the case of rhombic dodecahedron cell structures, the flow stress of the transverse direction was higher than that of longitudinal direction. Oscillations of the stress–strain curve shown in the transverse direction were due to the macroscopic shear band formation. Experimental results for the present aluminum foams were similar to the previous results for titanium foams [10]. High initial peak stress shown in a bcc-Voronoi cell specimen is probably due to the shorter cell edge length of the truncated octahedron cell.

Porous aluminum alloy specimen of fcc-Voronoi [001] cell structure with 91% porosity showed no oscillation of the stress–strain curve. There is no macroscopic

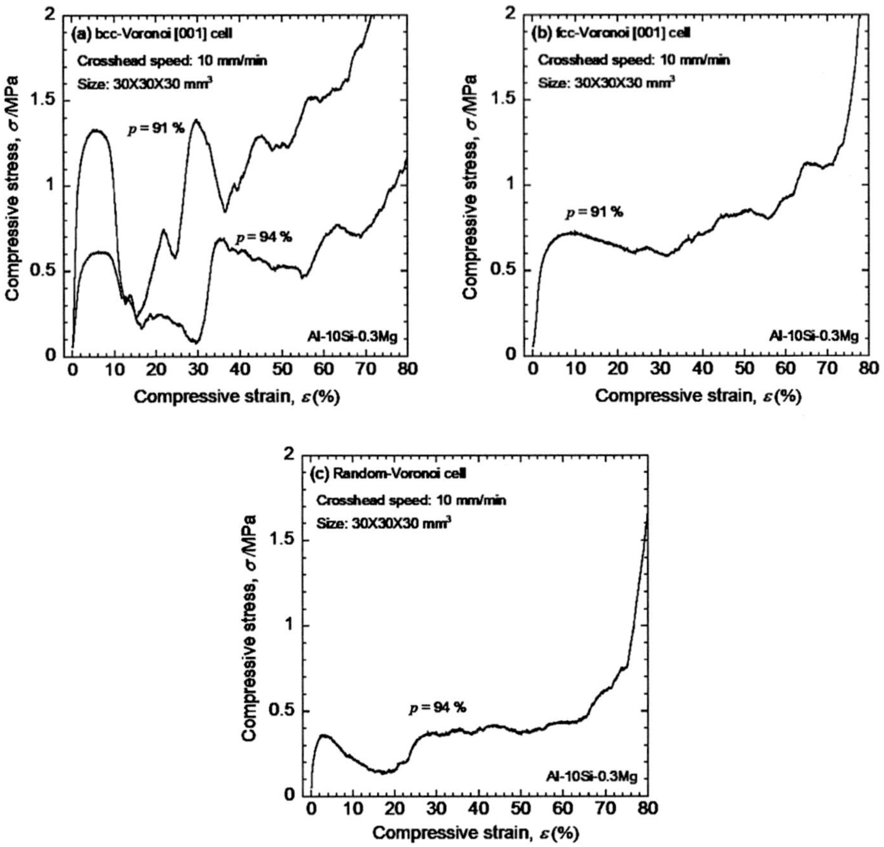

Fig. 5 Compressive stress–strain curves of porous aluminum alloy specimens with **a** bcc-Voronoi [001] cell, **b** fcc-Voronoi [001] cell, and **c** random Voronoi cell

shear band formation during the compression. However, the flow stress of fcc-Voronoi cell structure was smaller than that of bcc-Voronoi cell structure.

In the case of random Voronoi cell structure with 94% porosity, no oscillation was observed in the stress–strain curve. However, the flow stress was quite low compared to the bcc-Voronoi cell structure with 94% porosity. Low flow stress in the random Voronoi cell structure is due to the stress concentration by heterogeneous cell volumes.

Application to Spacecraft

Japan Aerospace Exploration Agency, JAXA, has a project for smart lander for investigating moon, SLIM. The purpose of the SLIM project is to pinpoint landing

Fig. 6 Images of the JAXA smart lander [11]

on the moon using a small spacecraft. Image of the spacecraft is shown in Fig. 6 [11]. Five AM porous aluminum alloy parts with acorn shape are used as energy-absorbing system of the landing. Investigation on the effect of compression speed on the energy-absorbing properties is in progress.

Conclusions

Porous Al–10Si–0.3 Mg alloys for space crafts were manufactured through AM process. Ordered and disordered cell structures were designed by 3D-CAD software based on 3D-Voronoi division. Compression tests using cubic specimens revealed that the ordered bcc-Voronoi cell structures have relatively high initial peak stress and absorbed energy. However, the deformation processes showed macroscopic shear band formation, which causes the reduction of flow stress. On the other hand, the ordered fcc-Voronoi and disordered random 3D-Voronoi specimens showed relatively high energy absorption efficiency at the expense of the compressive strength. It is considered that the partially disordered cell structures may have a potential to energy-absorbing applications compared to the fully ordered and disordered cell structures.

Acknowledgements This study was financially supported in part by SLIM project, Japan Aerospace Exploration Agency, and The Light Metal Educational Foundation, Japan.

References

1. Banhart J (2001) Manufacture, characterization and application of cellular metals and metal foams. Prog Mater Sci 46:559–632. https://doi.org/10.1016/S0079-6425(00)00002-5
2. Baumeister J, Banhart J, Weber M (1997) Aluminium foams for transport industry. Mater Des 18:217–220. https://doi.org/10.1016/S0261-3069(97)00050-2

3. Kitazono K, Tada R, Sugiyama Y, Miura T (2018) Impact energy absorbing system for space lander using hemispherical open-cell porous aluminum. Mater Sci Forum 933:337–341. https://doi.org/10.4028/www.scientific.net/MSF.933.337
4. Attar H, Calin M, Zhang LC, Scudino S, Eckert J (2014) Manufacture by selective laser melting and mechanical behavior of commercially pure titanium. Mater Sci Eng A 593:170–177. https://doi.org/10.1016/j.msea.2013.11.038
5. Leary M, Mazur M, Elambasseril J, McMillan M, Chirent T, Sun Y, Qian M, Easton M, Brandt M (2016) Selective laser melting (SLM) of AlSi12Mg lattice structures. Mater Des 98:344–357. https://doi.org/10.1016/j.matdes.2016.02.127
6. Rashed MG, Ashraf M, Mines RAW, Hazell PJ (2016) Metallic microlattice materials: a current state of the art on manufacturing, mechanical properties and applications. Mater Des 95:518–533. https://doi.org/10.1016/j.matdes.2016.01.146
7. Körner C, Singer RF (2000) Processing of metal foams—Challenges and opportunities. Adv Eng Mater 2:159–165
8. Moukarzel C (1993) Voronoi foams. Phys A 199:19–30. https://doi.org/10.1016/0378-4371(93)90093-J
9. Ashby MF (1983) The mechanical properties of cellular solids. Metall Trans A 14:1755–1769. https://doi.org/10.1007/BF02645546
10. Yue X, Matsuo K, Kitazono K (2017) Compressive behavior of open-cell titanium foams with different unit cell geometries. Mater Trans 58:1587–1592. https://doi.org/10.2320/matertrans.L-M2017834
11. Namikoshi H, Sugawara H, Chiba A, Washino R, Shuto K, Shimizu Y, Shibasaki Y, Shimoji H, Sato K, Yokoi T, Kono J, Miuki D, Aoki Y, Sakai S, Kushiki K, Sawai S, Fukuda S, Ishida T (2018) Preliminary system design of SLIM. In: Proceedings of 62nd space sciences and technology conference JSASS-2018-4081

Part II
Fabrication

Sonication for the Porosity Gradation of Foams Meets Replica Templating: A Hybrid Manufacturing Process for Lightweight Multifunctional Structures

C. Torres-Sánchez

Abstract Advancements in the design of porous structures with high-performance properties offer opportunities to many industries such as transport (towards diminished fuel emissions via mass reduction) or medical devices (achieving personalised orthopaedics by mimicking bone). However, the manufacturing of those parts lags behind because it is difficult to realise a complex porous structure design via conventional routes. Traditional manufacturing methods (e.g. casting) cannot produce complex porous distributions, and new ones (e.g. 3D printing) have a limited palette of processable materials. We propose a hybrid manufacturing process in which a sacrificial cellular structure (i.e. a polymeric porous foam) is tailored using a sonication technology that increases density in zones where reinforcement is needed (e.g. stress loading) and decreases it where mass is not needed, achieving lightweight topologically optimised structures. This template is then used in a replica process to create porous metal structures with an ad hoc porosity distribution for a specific application. Their characterisation required the creation of a bespoke tool that could combine complex geometry, porosity distribution, and properties at a given location simultaneously. The multifunctionality is achieved by coupling features, thanks to the gradation within the structure which grants enhanced coexisting thermal, shock-absorbent, controlled-diffusing, and/or vibration-damping properties. These lightweight multifunctional structures are unmatched by those fabricated by conventional production methods.

Keywords Sonication · Replica templating · Porosity gradation · Multifunctional · Polyurethane foam · Titanium alloy · Characterisation

---

C. Torres-Sánchez (✉)
Wolfson School of Mechanical, Electrical and Manufacturing Engineering, Loughborough University, Loughborough LE11 3TU, UK
e-mail: c.torres@lboro.ac.uk

N. Dukhan (ed.), *Proceedings of the 11th International Conference on Porous Metals and Metallic Foams (MetFoam 2019)*, The Minerals, Metals & Materials Series, https://doi.org/10.1007/978-3-030-42798-6_2

Introduction

Many mechanical and structural engineering applications require porosity-tailored structures. For example, one of the approaches pursued by the transport industry to curb CO_2 and NO_x emissions is to use lightweight materials. Because of the limits implicit in refining engine efficiency and performance, the final leap needs to be aided by substantially reducing vehicle mass, approximately 25% [1]. Another application where porous structures are desired is in tissue engineering: in order to achieve personalised implants, porous scaffolds need to be designed and manufactured so their macrostructure and mechanical properties resemble those of human bone in load-bearing anatomical locations, minimising *stress-shielding* and maximising stress redistribution, which derive on a stable bone cell/implant interface for a long-term fixation [2].

Lightweight porosity-graded structures can be achieved by reducing mass from the zones not subjected to the loads and stresses when in service, or by reinforcing those areas that are most exposed and at risk of failure by selectively introducing more mass in the matrix. The resultant material is a porosity-tailored structure with selectively reinforced areas that outperforms its solid counterpart and at a fraction of the weight. Topology optimisation techniques have made great strides in the pursuit of the design of these lightweight structures [3]. Despite the current demand for these materials with a porosity gradient and the work undertaken in modelling, the development of suitable manufacturing processes which would pave the way towards *mass customisation* lags behind. This is because there are enormous difficulties in fabricating such structures using traditional methods; physically joining sections of different porosities fail at the discontinuities, are highly labour-demanding, and difficult to adapt to automation for mass customisation.

Our proposed method uses a hybrid manufacturing process that combines a *Sonication technique* for the porosity gradation in a sacrificial polymeric foam and the *Replica templating* which employs this polymeric foam as the template structure for a metal slurry that will produce a metal foam after a thermal process.

Sonication Technique

Since the metal structure will copy the skeleton of the polymeric foam's, tailoring the latter is crucial. The manufacturing strategy used to modify the polymeric matrix is based on a controlled and timely irradiation of a beam of ultrasound directed onto a porous material while it is forming, i.e. foaming. This needs to occur during the 'sonication window' of the polymeric reaction, when the sonication energy couples with the gas bubbles embedded in the polymeric matrix and alters the final dimensions of the bubbles by sustaining them to a cyclic expansion/contraction due to the sinusoidal nature of the soundwave field. Under conditions of stable cavitation, the expansions are more extensive than the contractions, so the bubble growth is in

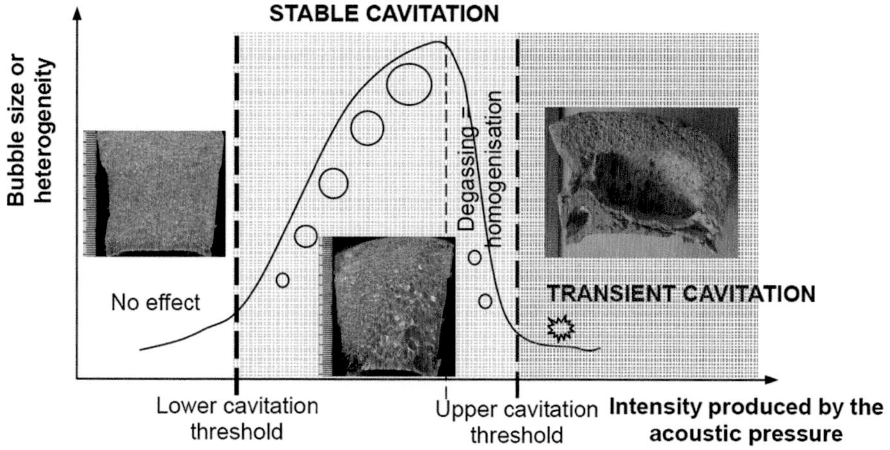

Fig. 1 Effect of stable cavitation on pore size depicting lower and upper thresholds of intensity, modified from [5]

resonance with the soundwave and sustained over time. The gas bubbles created (or added) and sonicated during the polymeric reaction will become the void sites of the porous material once cured, and their size will dictate the pore size and porosity, with larger or smaller pores where the matrix was stretched or shrunk by the ultrasound.

Stable cavitation induced in reacting polymeric foams can be achieved at different intensity levels. Below a lower threshold of intensity, sonication has no effect on the cellular structure, and beyond an upper threshold provokes degassing of the polymer melt and to a larger extent imploding of the bubbles and collapse of the foam (i.e. 'transient cavitation' [4]). Between these two thresholds, the range of sonication intensity favours stable pulsation of the bubbles in the viscoelastic matrix, and the porosity of the samples is found to increase with the intensity of the signal [5] (Fig. 1).

Replica Templating

The replica templating method is based on Schwartzwalder and Somer's patent in 1963 [6] for sintered refractory ceramics and has been applied to the fabrication of Ti alloys [7]. This technique uses a foamed polymer as a pre-form for the porous material. The polymeric foam is dipped into a slurry containing a binder and the metal powder that will form the final porous material. The soaked polymer is then heat-treated. At high temperatures, the polymer decomposes and a green body with similar porous structure to the pre-form polymer is obtained. They are typically open-cell structures [8], which are ideal for applications where open channels are required (e.g. multifunctional structures with thermal, filtering, catalytic, acoustic, flow transport properties). One of the advantages over other manufacturing processes is the possibility of fabrication without increasing pressure conditions, which places this method ahead of some other more energy-hungry procedures. One drawback is that the pyrolysis of the polymer skeleton in many cases produces flaws or leaves residues in the cellular structure, and this can be detrimental to the mechanical properties of the final foam [9, 10]. In addition, some dimensional shrinkage could occur during fabrication if the concentration of the slurry is not well defined [11].

Characterisation Methods

The characterisation of the porosity distribution in these metal foams is of vital importance because these data can be used to feed back/feed forward the manufacture process and further improve their properties and functionalities according to the topology-optimised model. Traditional density measurement techniques (e.g. liquid displacement techniques such as Archimedes's, Helium infiltration displacement, mercury porosimetry, etc.) provide information on the bulk values only and disregard or cannot report on location (i.e. do not link density scalar values with a specific location), which is frequently critical when mechanical properties or other functionalities have to be engineered. A standard methodology for the characterisation and quantification of porosity gradient materials is still missing.

This paper presents a novel hybrid manufacturing process that created porosity gradation within a titanium alloy foam using a template (i.e. polyurethane foam) that had been previously modified with Sonication. After the slurry-infiltrated pre-form had been sintered, mechanical testing was performed on homogeneous zones because the overall characterisation of a porosity-graded structure is not possible via standard methods. Consequently, we propose a method that maps porosity to location and to properties, and paves the way towards parameterising property-gradient cellular structure foams in an attempt to accelerate fabrication and deployment of these structures in a myriad of industrial applications and consumer products.

Methodology

Materials

The sacrificial material chosen for this study was a polyurethane foam and the metal powder was a ternary Titanium (Ti) alloy with Niobium (Nb) and Tin (Sn) as alloying elements. Ti alloys in general are sought-after materials in the aircraft, aerospace, and niche automotive industries due to their low density (compared to stainless steel), high corrosion resistance, and toughness. This ternary alloy has been reported in literature for its promising performance as a medical device in bioengineering and tissue replacement applications [12, 13].

Polyurethane foams were made by the polymerisation of a monomer mixture (polyol and isocyanate) after the addition of a catalyst and a blowing agent (e.g. water), and the urethane groups created CO_2 gas as a by-product of the reaction. This frothing process follows five characteristic stages: cream, rise, packing, and gelation, followed by curing, or solidification, of the foam [14–16]. The rising and packing stages in polyurethane have been reported as the 'sonication window' for this application [17, 18].

Elemental powders of Ti (Alfa Aesar, MA, 99.5% purity, $\leq$45 μm, –325 mesh), Nb (Aldrich, GE, 99.8% purity, <45 μm, –325 mesh), and Sn (Alfa Aesar, MA, 99.8% purity, $\leq$45 μm, –325 mesh) were blended together according to the desired composition of Ti 61, Nb 35, and Sn 4 wt%. This alloy is referred to as Ti35Nb4Sn thereafter. Mechanical alloying was performed in a planetary ball mill (Pulverisette 6 Monomill, Fritsch, Germany) under control of argon atmosphere using stainless steel grinding bowl and balls (10 mm diam). The ball-to-powder weight ratio was 10:1. The ball milling was carried out at room temperature and a rotation speed of 200 rpm for 12 h. To prevent cold welding of the metal particles to the surface of the bowl and balls, 2 wt% stearic acid ($CH_3(CH_2)_{16}COOH$) was added as a lubricant. Phase constituents in the ball-milled powders and in the sintered porous scaffolds were characterised using X-ray diffractometry (AXS D8Advance glancing, Bruker, Germany) with 1.5046 nm wavelength Copper K-alpha radiation source scanned using 0.009° per step and 46.2 s as step time. The foams were examined using a scanning electron microscope (Quanta 3D FEG dual beam, 3.5 nm, 30 kV, DCG Systems Inc., USA) and a microfocus tomography system (Skyscan m-CT, Bruker, Belgium) with a 225 kV X-ray head with a 225 W target. The reconstruction software (X-Tek Systems Ltd, Group, UK) was used to produce volumetric images.

Sonication

A measured amount of reactants (2 ml polyol and 2 ml isocyanate) were poured into a plastic container (PP, 1.2 mm thickness). The process was initiated by the addition of distilled water (i.e. the chemical blowing agent) that acted as the catalyst to the reaction. After stick-stirring for approximately 5 s, the container was placed in a temperature-controlled water bath and located at 11.1 cm from the sonication probe irradiating at a frequency of 20 kHz and 150 W power. Full details of the procedure have been described in [17]. On completion of the reaction and sonication, the foams were left to cure for 48 h, de-moulded, and used without further modification.

Replica Templating and Sintering

The slurry was prepared by mixing 75 wt% of Ti35Nb4Sn powders with a binder (25 vol% ethanol/5 wt% polyvinyl alcohol mixture in distilled water). The slurry was stirred continuously to minimise sedimentation. The polyurethane foams were dipped into the slurry for 24 h using a pressurised chamber with a pump that applied a light negative pressure to induce uptake of the slurry into the polymeric foam. The soaked foam-metal green bodies were then placed in an atmosphere-controlled (Argon) furnace (Lenton Thermal Designs, UK). The sintering process was conducted in two stages: first, the green bodies were heated at 2 °C/min from room temperature up to 250 °C, well beyond the melting point range reported for polyurethanes, and kept at that temperature for 2 h to sublimate the polymeric skeleton that left a fragile metal 'wire' structure behind. During this stage, the furnace fumes were continuously evacuated. Following that, the furnace was flushed three times to remove gases resulting from the decomposition of the polyurethane that could oxide the metal foams. A second step followed and the 'wire' structures were sintered at 1,100 °C (heating rate of 5 °C/min) and held at this temperature for 3 h. The furnace was then allowed to cool down to room temperature. Samples were cleaned using acetone and soapy water in an ultrasonic bath for 5 min and left to air dry.

Mechanical Properties: Compression Testing

The foams were sectioned into cubic specimens that presented an isotropic homogeneous texture. These sections were used for mechanical characterisation. The compression experiments were conducted at room temperature in a Universal Testing Machine (3367 Instron, UK). A compressive rate of 1 mm/min was used on the samples until plastic deformation and/or destruction. Stiffness was calculated from the gradient created between compressive strength and compressive strain in elastic conditions. Compressive yield stress was obtained from the 0.2% strain offset beyond the transition 'point' at which the material started its plastic behaviour.

Porosity Distribution Characterisation

The exploitation of porosity-graded foams requires a robust method to characterise the complex geometry associated with their cellular structure and local properties. The development of the 'Topo-porosity' tool [18] was motivated by this need. Since porosity is defined as the fraction of the total volume which is empty, bulk density is directly related to the volume fraction of the solid phase of a foam. In the context of a 2D image, bulk density is equivalent to area fraction of the solid phase. In image analysis, identifying pore boundaries through edge finding and network analysis is very difficult given the varying geometry of a foam, but measuring the volume of the material present (or absent) in a region is relatively easy. In a graded foam, the amount of mass distribution varies within different area fractions, and this value, relative to an ideal homogeneous distribution of porosity, is the variable of interest.

After fabrication, the foams were cut into half along their axial plane to expose the cross section of the foam orthogonal to the direction of propagation of the ultrasound. Samples of foams were scanned at a 1500 dpi resolution on a 1640SU scanner (EPSON Perfection, Seiko Group, Japan) and these images processed in the 'Topo-porosity' tool. A grid was applied to the image, and each grid square associated with a value which was the number of pixels contained in that area corresponding to the solid phase. The square size was adapted (i.e. reduced or expanded) so it contained the number of pixels given by the average value of the entire image. A *contour* function was used to generate a set of curves, i.e. isolines, which linked squares of equal size (i.e. area fraction) in the same way that contour lines in a topographic map connect continuous points of the same altitude. These *topographic maps of density* provide information on the porosity distribution within a foam's cross section, indicating relative positions of areas with an equivalent density.

Results

The polyurethane foams produced by sonication displayed a degree of porosity gradation in their cross section, as presented in Fig. 2a, and an open-cell structure (Fig. 2b). These foams were then used to obtain metal foams by 'replica' templating (Fig. 2c). Volumetric reconstructions (Fig. 2d) were created from the micro-CT images obtained from the metal foams, which allowed the observation of a radial porosity gradation in the example presented in Fig. 2e. The characterisation of the constituent elements of the Ti alloy before and after sintering is presented in Fig. 3.

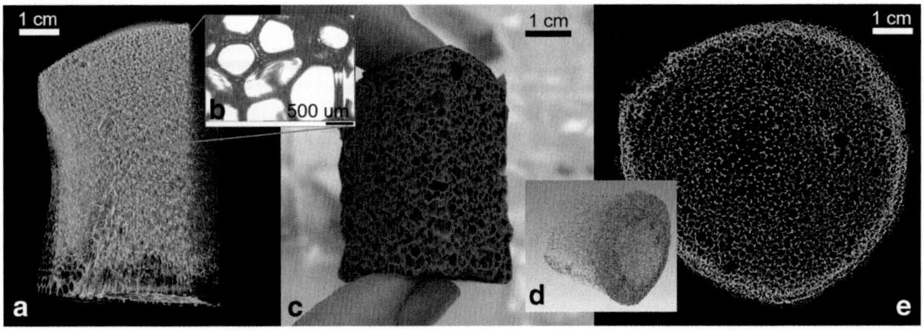

Fig. 2 **a** μCT of sacrificial PU foam; **b** SEM of PU foam displaying the reticulated structure of struts; **c** Metal foam; **d** 3D reconstructed volume of the metal foam; **e** μCT cross section of the metal foam depicting radial porosity gradation

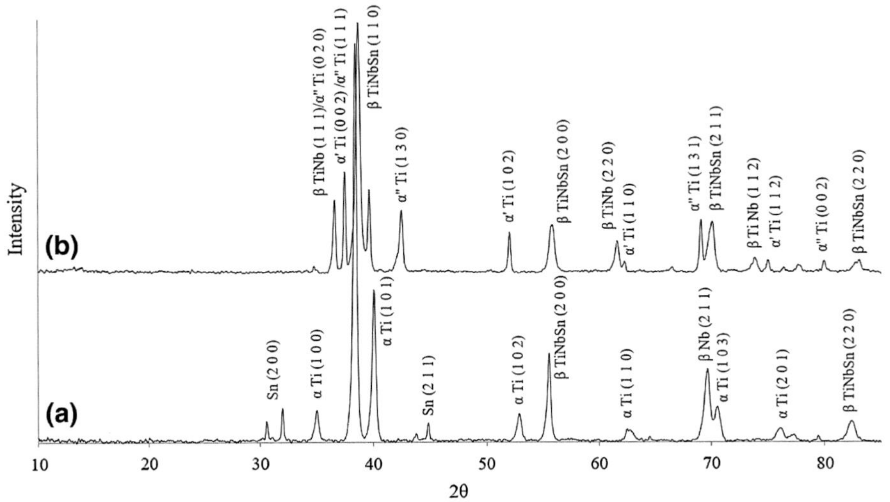

Fig. 3 XRD patterns acquired for **a** ball-milled powders for 12 h; **b** after sintering for 3 h

The results of stiffness (i.e. E in MPa) for each of the porosities (evaluated as volumetric ratios) are presented in Fig. 4. Yield strengths were plotted against stiffness to study the decrease of one versus the other (Fig. 5) and how they fit to a linear trend. Two batches were used for this purpose to expand the range of porosities obtained via this hybrid manufacturing process.

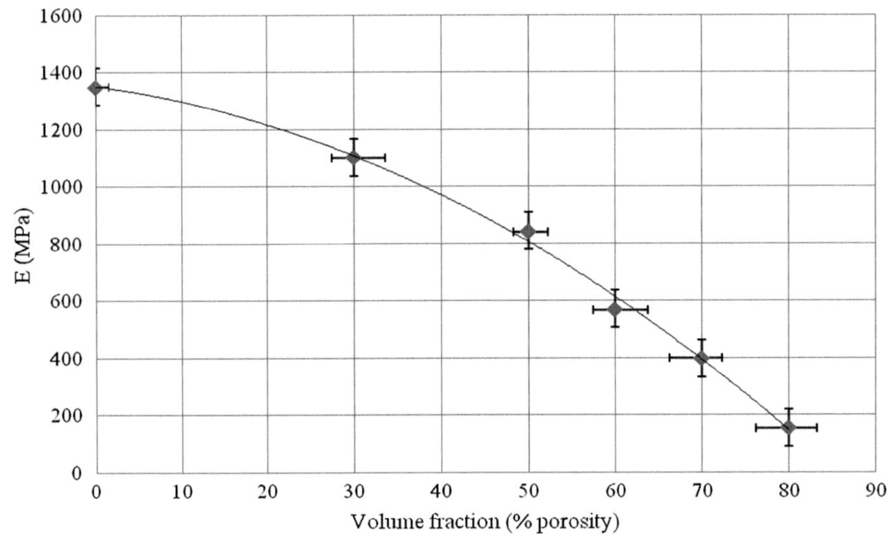

Fig. 4 Stiffness value versus vol% porosity (standard deviation ranging 3–11%) for the Ti35Nb4Sn metal foams

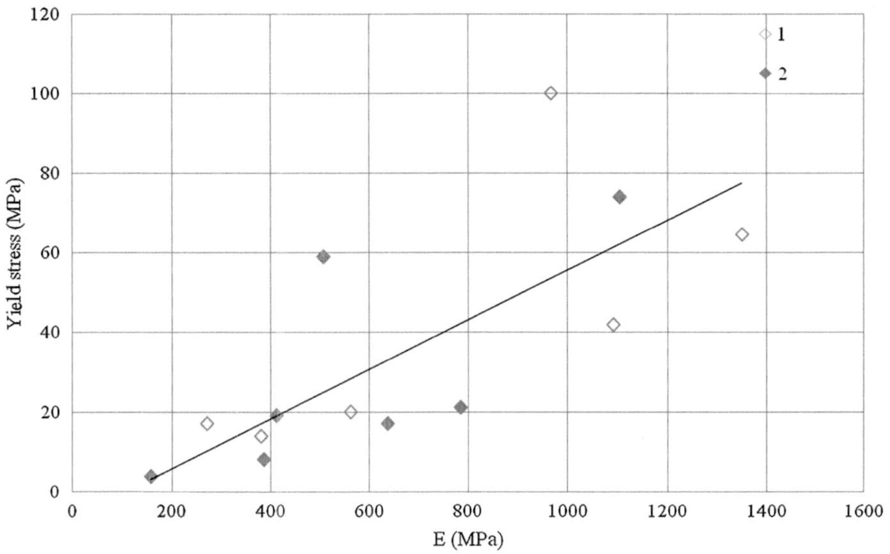

Fig. 5 Yield stress versus stiffness for the Ti35Nb4Sn metal foams

The results from the 'Topo-porosity' tool allowed a quantification of the porosity gradation. The contour maps displayed values of vol% density across the cross section, and it was possible to couple values of porosity with location (Fig. 6).

Discussion

The porosity gradation incurred in the sacrificial PU foam by using the Sonication technique permitted the generation of a final metal foam with a graded macrostructure, a replica of the PU skeleton after removal of the template and sintering of the powders. The analysis of the constituent elements of the ball-milled Ti35Nb4Sn powders (Fig. 3a) shows that the alloying of the three elements present in the mixture was achieved to a certain extent. The formation of beta TiNbSn (cubic) phase was achieved after the 12 h ball milling process and stabilised at room temperature by the complete dissolution of Nb and Sn into the Ti beta phase. Some Sn (originally tetragonal), Nb (originally cubic, beta), and Ti (originally hexagonal, alpha) were still present in small proportions in the XRD pattern of the ball-milled powders. On the contrary, by the end of the ball milling process, the stearic acid did not show on the XRD pattern. Figure 3b shows the XRD pattern of the sintered samples. The dominant beta (BCC) phase coexisted with alpha (hexagonal) and alpha" (orthorhombic) phases. The presence of alpha phases was expected because, although a large amount of Nb present in the elemental mixture would stabilise the beta phase, the sintering energy was insufficient to achieve a full stabilisation. The alpha phase could be

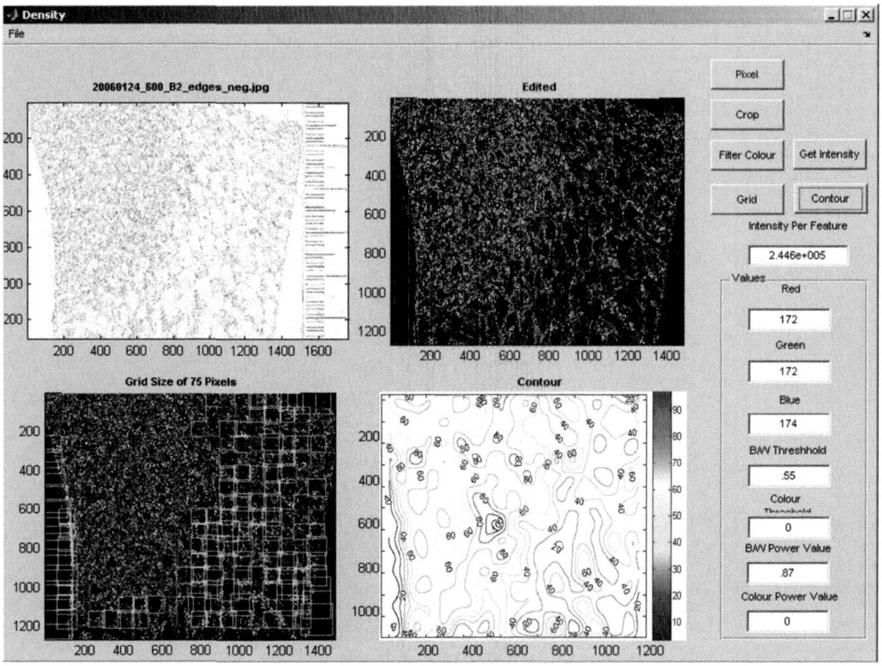

Fig. 6 Topo-porosity interface during the analysis of a metal foam. The output is the vol% density contour map in the bottom right corner

explained by an incomplete alloying of the elements or the presence of Ti oxides. The presence of TiNb (FCC) was also an indication of only partial alloying achieved during sintering or oxidation. The stiffness values obtained from the compression testing of the foams fell in the range of 200 MPa–1.4 GPa for porosity range from 80% to quasi-solid, respectively. The stiffness values at around 70–80 vol% porosity (i.e. 430–100 MPa, respectively) were lower than those reported in literature (i.e. E ranging 430–500 MPa at 75% porosity) [8], and E ~ 580 MPa at ~84% porosity [] but similar to [20] (i.e. E ~ 300 MPa at ~75 vol% porosity) and [10] who also used a very aqueous so19lution for the slurry. This could be a manifestation of the unfinished alloying and the oxides present in the structures, which prevented the powder particles from sintering effectively [12]. The corresponding strength values spanned from 3 to 100 MPa, again lower than those reported for ~75% porosity as ~24 MPa [20] but similar to those of highly porous foams (e.g. ~80% porosity, yield strength 15 MPa [9]; 84% porosity, yield strength 10 MPa [19]; 88% porosity, yield strength 7–13 MPa [7]). The stiffness of the foams and their compressive strengths correlated relatively well, in agreement with other values reported [21–23]; however, the trendline gradient (Fig. 5) was smaller than expected because the pace at which the stiffness increased with density was not matched with that of strength. The manufacturing process reported here did not include the use of plasticisers, unlike [8], or dispersant agents, used by [19], but only a binder. The process involved a single slurry-infiltration step, unlike [24, 25] who studied the effect of multiple cycles to improve mechanical strength. These process parameters could explain why the results obtained are lower than those reported elsewhere and that the prevalent failure mode was brittle microfracture, a symptom of unwanted oxygen contamination.

Stiffness and density (i.e. 100 vol% porosity) ratio was fitted to a power law relationship ($y = x^n$), and the n index ranged between 1.30 and 1.36 which corresponds to an interconnected porous structure with a dominant network of channels in an open architecture [26], confirming the hypothesis that the metal foam would replicate the macrostructure of the sacrificial PU foam. If only high porosities are considered (i.e. >50%) a linear relationship can be established, so n becomes 1, manifestation of an architecture that is fully open and interconnected.

But this is only applicable to sections of homogeneous porosity and does not consider porosity gradation within the entire sample volume. This is a major drawback when it comes to characterising graded materials that were designed to display multifunctionality or multiphase distributions (e.g. via topology optimisation of high density/high porosity zones). Motivated by the requirement of a procedure to parameterise this gradation, a tool for characterising porosity distribution, i.e. *'Topo-porosity'*, has been devised for the mapping of the distribution of porosity within a graded cellular structure. The automated measurement protocol allows area characterisations that established methods cannot provide.

Validation of the Method

One of the research questions that arose regarding the validity of the 'Topo-porosity' mapping tool was its sensitivity at the boundaries between areas of very different densities, common scenario in porosity-tailored foams. To test this, a series of digitally stitched images were created joining pictures of homogeneous foams of known porosities (e.g. very low porosity area 1 joined to a high porosity area 2, in ascending order (Fig. 7a) and tiled (Fig. 7b). The results demonstrated that this method could detect both incremental (Fig. 7c) and abrupt (Fig. 7d) changes in porosity, reported the values in vol% density, and these were represented in the corresponding contour maps.

As can be seen in Fig. 8, there is a direct correlation between the density contour maps (Fig. 8b) and the visual density distribution in the foams' pictures. Overlaid onto that was the stiffness data (Fig. 8a) in an attempt to characterise porosity-graded foams in terms of their mechanical performance coupled with location within the structure. This is the first step towards volumetric quantification of porosity and mechanical properties, a more data-rich proposition compared to traditional bulk properties that are only valid for homogeneous materials.

The advantages of this hybrid method that combines *Sonication and Replica templating* are the following: (i) compared to 'cut-and-paste' traditional methods, this is a flexible way to engineer porosity across a single cross section; (ii) as the metal is in solid state at all stages, it minimises issues with oxidation and health and safety

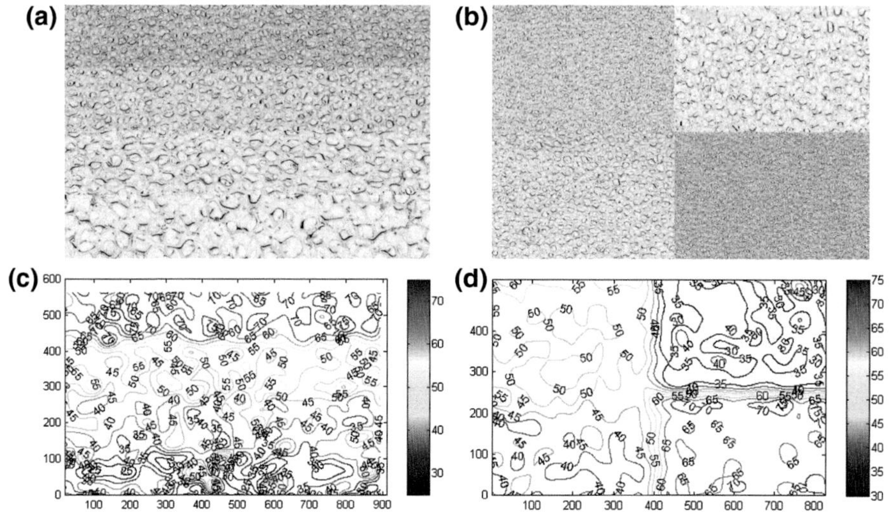

Fig. 7 Validation exercises with (**a** and **c**) incremental changes in density and (**b** and **d**) abrupt changes to demonstrate the sensitivity of the *Topo-porosity* tool

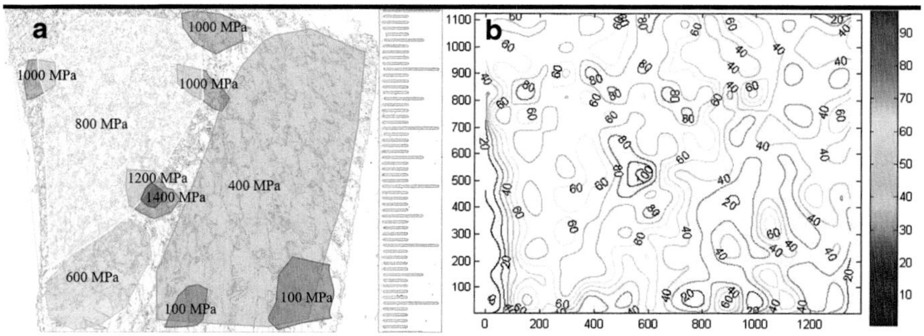

Fig. 8 **a** Stiffness values overlaying a cross-sectional scan to depict local variation of mechanical properties and **b** topo-porosity results reporting vol% density in a porosity-graded metal foam

for the operators; (iii) it is cost-effective as it uses versatile starting materials, such as polyurethane foams; (iv) it proposes a step towards mass customisation, with shape and form freedom and in low volume production batches; (v) it presents potential for automation; and (vi) it can create multifunctional and tailored structures and alloys.

Conclusions

New structures and manufacturing protocols to design and realise engineered structures that combine high-spec mechanical properties and low weight are sought after. Ideally, one would prefer a method that creates a gradient in the porosity within the bulk of the material without the need of manual finishing or post-processing, avoids the use of expensive or hazardous materials, is cost-effective, and has a potential for automation. We propose a hybrid manufacturing process for lightweight multifunctional structures using a *Sonication technique* for the porosity gradation of a sacrificial polymeric foam which then is used as a template for the metal foam via the *Replica process*. To assist their fabrication, a suitable characterisation method that could handle heterogeneity of the graded structure had to be devised. Unlike other measurement methods where statistics of features from one pixel carry no information on its location, this method brings together both types of information, porosity (or density), ratio, and location. Overlaying properties (e.g. mechanical) to these maps achieves a multidimensional dataset that takes manufacturers a step closer to the realisation of these topology-optimised structures.

Acknowledgements The author gratefully acknowledges funding received from EPSRC (grant ref. EP/P027482/1) and from Innovate UK (grant ref IDP14/7724). Undergraduate students Craig Maxwell, Daniel Wo, John McLaughlin, and Alan Cheater are acknowledged for their contributions while using the techniques presented in this paper during their projects or dissertations.

References

1. The International Council on Clean Transportation (ICCT) (2018) Vehicle mass reduction (lightweighting), 1 March 2018
2. Torres-Sanchez C, Al Mushref FRA, Norrito M, Yendall K, Liu Y, Conway PP (2017) The effect of pore size and porosity on mechanical properties and biological response of porous titanium scaffolds. Mater Sci Eng: C 77
3. Rozvany GIN (2009) A critical review of established methods of structural topology optimization. Struct Multidiscip Optim 37:217
4. Gallego-Juarez JA, Rodriguez-Corral G, Riera-Franco de Sarabia E, Vazquez-Martinez F, Acosta-Aparicio VM, Campos-Pozuelo C (2001) Development of industrial models of high-power stepped-plate sonic and ultrasonic transducers for use in fluids. In: Proceedings of the IEEE ultrasonics symposium, pp 571–578
5. Torres-Sanchez C, Corney JR (2009) Toward functionally graded cellular microstructures. J Mech Des 131(9)
6. Schwarzwalder K, Somers AV (1963) Method of making a porous shape of sintered refractory ceramic articles, US3090094
7. Li JP, Van Blitterswijk CA, de Groot K (2004) Factors having influence on the rheological properties of Ti6A14V slurry. J Mater Sci: Mater Med 15(9)
8. Hsu H-C, Hsu S-K, Wu S-C, Wang P-H, Ho W-F (2013) Design and characterization of highly porous titanium foams with bioactive surface sintering in air. J Alloys Compd 575
9. Ahmad S, Muhamad N, Muchtar A, Sahari J, Ibrahim MHI, Jamaludin KR, Nor NHM (2010) Development and characterization of titanium alloy foams. Int J Mech Mater Eng 5(2)

10. Tange M, Manonukul A, Srikudvien P (2015) The effects of organic template and thickening agent on structure and mechanical properties of titanium foam fabricated by replica impregnation method. Mater Sci Eng: A 641
11. Nangrejo MR, Bao X, Edirisinghe MJ (2001) Silicon carbide–titanium carbide composite foams produced using a polymeric precursor. Int J Inorg Mater 3(1)
12. Torres-Sanchez C, McLaughlin J, Fotticchia A (2018) Porosity and pore size effect on the properties of sintered Ti35Nb4Sn alloy scaffolds and their suitability for tissue engineering applications. J Alloys Compd 731
13. Matsumoto H, Watanabe S, Hanada S (2005) Beta TiNbSn alloys with low Young's modulus and high strength. Mater Trans 46(5)
14. Marciano JH, Reboredo MM, Rojas AJ, Williams RJJ (1986) Integral-skin polyurethane foams. Polym Eng Sci 26(11)
15. Modesti M, Adriani V, Simioni F (2000) Chemical and physical blowing agents in structural polyurethane foams: simulation and characterization. Polym Eng Sci 40(9)
16. Torres-Sánchez C, Corney J (2009) Identification of formation stages in a polymeric foam customised by sonication via electrical resistivity measurements. J Polym Res 16(5)
17. Torres-Sánchez C, Corney J (2008) Effects of ultrasound on polymeric foam porosity. Ultrason Sonochemistry 15
18. Torres-Sánchez C, Corney J (2009) Porosity tailoring mechanisms in sonicated polymeric foams. IOP J Smart Mater Struct 18(10)
19. Manonukul A, Tange M, Srikudvien P, Denmud N, Wattanapornphan P (2014) Rheological properties of commercially pure titanium slurry for metallic foam production using replica impregnation method. Powder Technol 266
20. Cachinho SCP, Correia RN (2007) Titanium porous scaffolds from precursor powders: rheological optimization of TiH2 slurries. Powder Technol 178(2)
21. Ramakrishnan N, Arunachalam VS (1990) Effective elastic moduli of porous solids. J Mater Sci 25(9)
22. Boccaccini AR, Ondracek G, Mazilu P, Windelberg D (1993) On the effective Young's modulus of elasticity for porous materials: microstructure modelling and comparison between calculated and experimental values. J Mech Behav Mater 4(2):119
23. Torres-Sanchez C, McLaughlin J, Bonallo R (2018) Effect of pore size, morphology and orientation on the bulk stiffness of a porous Ti35Nb4Sn alloy. J Mater Eng Perform 27(6)
24. Wang C, Chen H, Zhu X, Xiao Z, Zhang K, Zhang X (2017) An improved polymeric sponge replication method for biomedical porous titanium scaffolds. Mater Sci Eng: C 70
25. Li JP, Li SH, de Groot K, Layrolle P (2002) Improvement of porous titanium with thicker struts. Key Eng Mater 240–242
26. Gibson LJ, Ashby MF (1997) Cellular solids. Structure and properties. Cambridge University Press, Cambridge

Development of Rotational Incremental Hammering Process for Porous Metals

Han Cui, Ryo Matsumoto and Hiroshi Utsunomiya

Abstract Porous metals have huge potential in various engineering applications, but their applications have been limited due to the lack of studies on plastic working and metal forming. In this study, a new forming process named rotational incremental hammering was developed. Inspired by the friction stir welding process, the forming punch was rotated at high speed while indenting the workpiece to introduce localized heating and more intense plastic deformations. Open-cell nickel foam plates were shaped into simple stair-shaped components with the proposed process and two conventional metal forming processes: incremental hammering process and friction stir incremental forming process. It was found that the forming result obtained with the developed process showed improvements when compared with the forming results obtained with conventional incremental hammering process or friction stir incremental forming process. In addition, a thin skin layer was also found on workpiece formed with the developed process, which is useful to achieve sandwich structures.

Keywords Nickel foam · Incremental forming · Friction stir · Forming parameters · Plastic deformation · Fast-prototyping

H. Cui
Department of Materials Science and Engineering, Stanford University, Stanford, CA 94305, USA
e-mail: hancui@stanford.edu

R. Matsumoto · H. Utsunomiya (✉)
Division of Materials and Manufacturing Science, Graduate School of Engineering, Osaka University, Suita 565-0871, Japan
e-mail: uts@mat.eng.osaka-u.ac.jp

R. Matsumoto
e-mail: ryo@mat.eng.osaka-u.ac.jp

N. Dukhan (ed.), *Proceedings of the 11th International Conference on Porous Metals and Metallic Foams (MetFoam 2019)*, The Minerals, Metals & Materials Series, https://doi.org/10.1007/978-3-030-42798-6_3

Introduction

Commonly known as metal foams, porous metals are metals with pores integrated into their structure. Owning to this characteristics, porous metals possess a number of properties that cannot be obtained with fully dense metals such as low densities, large surface area, excellent energy absorption capability, as previously reviewed [1, 2, 5]. These properties have made porous metals a good candidate to be used in various functional and structural applications such as filters, batteries, biomedical implants, and bearings [4].

Despite huge potentials in various functional and structural applications, application of porous metals is still limited because of their disadvantages such as poor weldability and difficulties in applying conventional cutting and machining techniques as those techniques would cause severe damage to the cellular structure of porous metals [1]. In addition, applications of plastic working and metal forming techniques on porous metals are rather limited, though they are highly demanded in rapid prototyping and massive productions of components. Because pores may close during the forming process, advantages of porous metals such as low density and high porosity may be lost. In addition, because conventional plasticity theories are based on the assumption of volume consistency, it is also difficult to design industrial metal forming process for porous metals with predictable properties of formed components, and there were no published studies on applying metal forming processes such as incremental forming and friction stir incremental forming to form porous metals because the dimensional accuracy of the products is not satisfactory.

In this study, in order to improve the dimensional accuracy of formed components, a new metal forming process called rotational incremental hammering was proposed and implemented to form the open-cell nickel porous metals. The component formed with the proposed process was compared with those formed with conventional incremental hammering process and friction stir incremental forming process with the same forming parameters.

Overview of Rotational Incremental Hammering Process

Figure 1 shows the schematic illustration and the forming sequence of the rotational incremental hammering process. For the conventional incremental hammering process, the workpiece is pressed locally by the forming tool and this elementary operation is repeated at a specific depth and interval of indentation until the desired shape is obtained [7]. For the rotational incremental hammering process, however, the conventional incremental hammering process is combined with features of friction stir welding that the forming tool is rotating at a high speed while pressing the workpiece. Despite severe plastic deformations caused by the indentation and the rotation of the forming tool, the friction between the tool and the workpiece heats the material locally, and the localized heating softens the material around the forming tool.

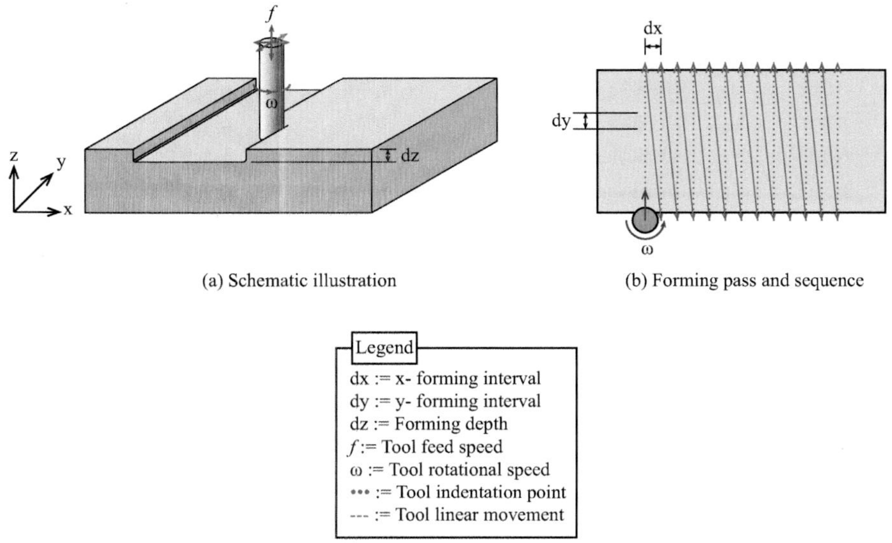

Fig. 1 Overview of rotational incremental hammering process

In addition, since cell walls near the surface of the workpiece are folded and overlapped by the combination of tool rotation and translation, a skin layer is fabricated on the surface of the workpiece and formed a sandwich structure, whose effects on the mechanical properties of various porous metals have been well studied [6, 8, 9].

Experimental Procedure

Materials

Open-cell nickel foams (CELMET #4 by Sumitomo Electric Industries, Ltd.) with mean pore size of 0.45 mm were used. The material was manufactured by adding electroconductive coating to a urethane foam with connecting pore, followed by electrolytic Ni planting and removal of urethane support [3]. Rectangular plates with dimensions of 60 mm (length) × 30 mm (width) × 10 mm (height) were prepared by wire electrical discharge machining as specimens, as shown in Fig. 2a. The specimens were then formed into stair-shaped components as shown in Fig. 2b. Table 1 summarized the main specifications of specimens used in this study.

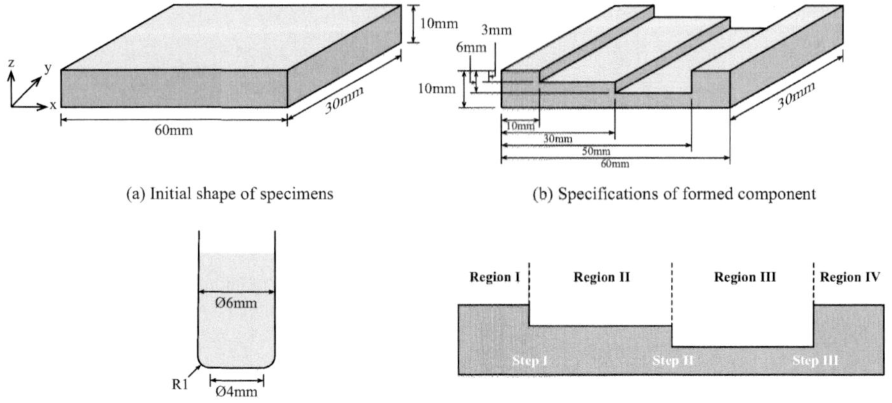

(a) Initial shape of specimens (b) Specifications of formed component

(c) Schematic illustration of the forming tool (d) Definition of terms "Region" and "Step"

Fig. 2 Experimental conditions

Table 1 Specifications of rectangular specimens

Length	60 mm
Width	30 mm
Height	10 mm
Mean pore size	0.45 mm
Average apparent density	$0.45 \pm 0.07 \mathrm{Mg\,m^{-3}}$

Forming Conditions

The forming experiments were carried out on a commercial four-axis CNC milling machine (Roland DGA Corporation MDX-540S). A rod-shaped tool made of SKH51 high-speed tool steel was used as the forming tool. The tool had a diameter of 6 mm and a flat end with a diameter of 4 mm, as shown in Fig. 2c. The specimen was held on a aluminum alloy die and fixed by ∅0.25 mm steel wires at both ends, as shown in Fig. 3. For comparison, the specimens were formed with both the proposed forming process, rotational incremental hammering (RIH), and two conventional metal forming processes, incremental hammering (IH), and friction stir incremental forming (FSIF).

For incremental hammering process, the forming tools without rotation locally press the specimen to cause plastic deformation, and the elementary operation is repeated over the entire specimen to obtain the desired component, as shown in Fig. 4a [7]. For the friction stir incremental forming process, which was originally developed by Otsu et al. as a novel metal sheet forming process, the specimen is pressed by the forming tool with a high rotational speed and plastically deformed by both the translational and rotational motions of the forming tool, as shown in Fig. 4b

Fig. 3 Experiment setup

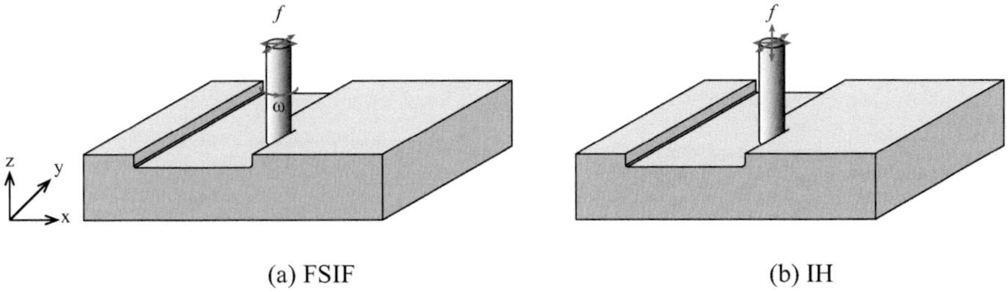

Fig. 4 Schematic illustration of **a** friction stir incremental forming process and **b** incremental hammering

[10]. Table 2 lists the forming conditions of IH, FSIF, and RIH process. Surface observations, cross-sectional observations, as well as laser displacement meters were employed to evaluate the forming results.

Results

Figure 5 shows the top-view and side-view images of formed specimens with the rotational incremental hammering (RIH) process, incremental hammering (IH) process, and friction stir incremental forming (FSIF) process.

Table 2 Summary of forming conditions of IH, FSIF, and RIH methods

Forming method	IH	FSIF	RIH
x-forming interval dx	1 mm	1 mm	1 mm
y-forming interval dy	1 mm	N/A	1 mm
Forming depth dz	1 mm	1 mm	1 mm
Tool rotational speed ω	0	8000 rpm	8000 rpm
Hammering speed f	60 mm/s	60 mm/s	60 mm/s

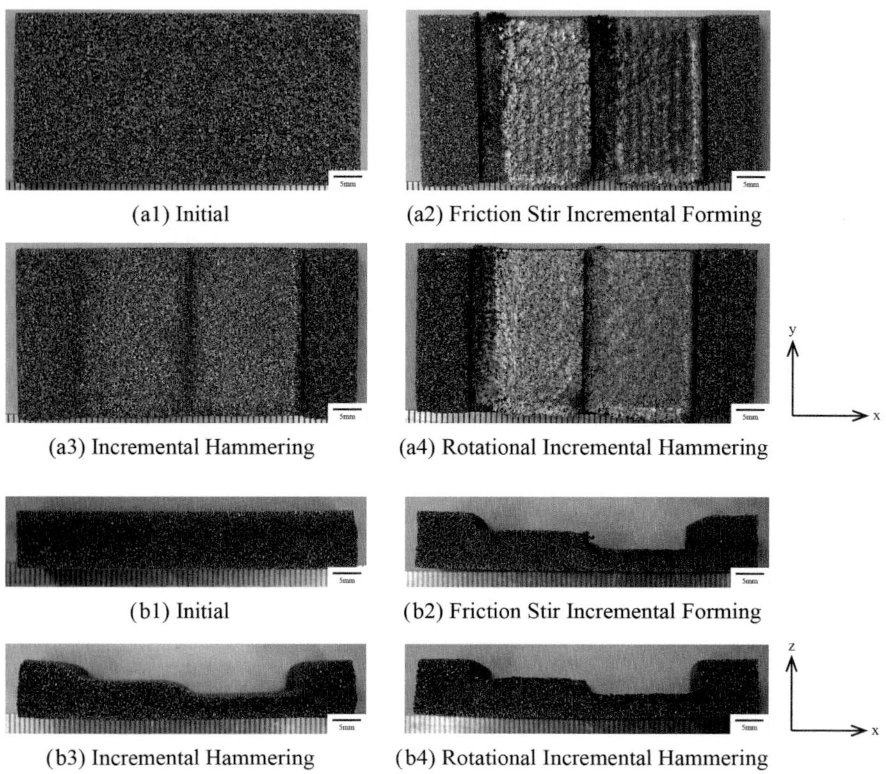

Fig. 5 Photographs of **a** top view and **b** side view of formed specimens with different forming processes

In addition, Fig. 6 shows the microscope images of the x-z cross section of the formed specimens, and Fig. 7 shows the surface profiles of the formed specimens measured with laser displacement meter.

For the specimen formed with the IH process, because the tool only hammered the specimen without rotation, the surface of the specimen was not stirred and only the cell walls collapsed, and no surface layer was formed. For the specimen formed with the FSIF process and the RIH process, however, a thin skin layer with thickness

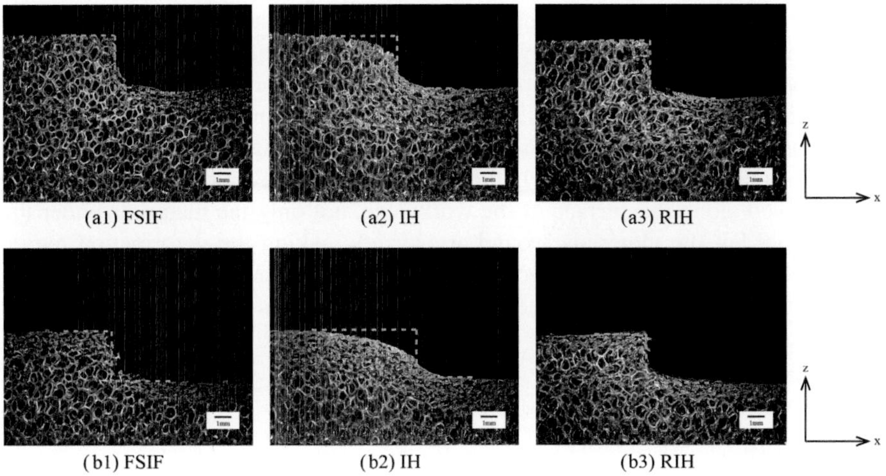

Fig. 6 Microscope images of **a** Step I and **b** Step II of formed specimens with different forming processes

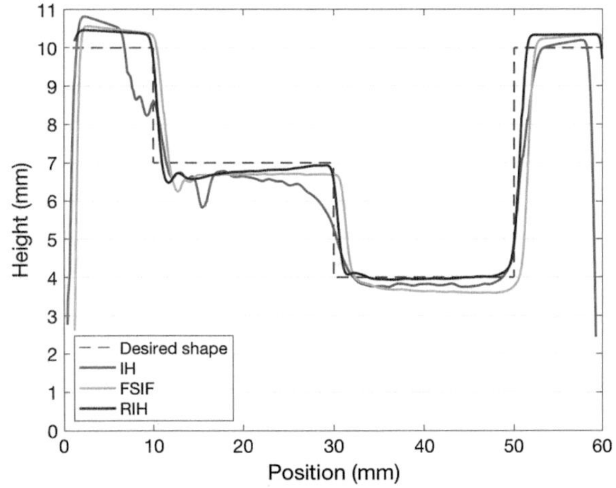

Fig. 7 Surface profiles of formed specimens with different forming processes

of approximate 1 mm (FSIF) and 0.75 mm was found on the surface of the specimen through SEM observation (images not shown), which is mainly due to folding and overlapping of cell walls resulted from the tool rotation.

Compared to the specimens formed with IH process, the shape accuracy of each "stair" on the specimen formed with the RIH process has significant improvements, as shown in Fig. 6. Because the tool is rotating at high speed while hammering the specimen, the temperature of the workpiece increases because of friction between the forming tool and the workpiece and plastic deformation of the workpiece. As a result of the localized heating, the materials around the forming are softened, causing more severe plastic deformations.

Moreover, from Fig. 7, although the shape of specimens formed with all three forming methods did not perfectly match with expectation, specimen formed with the RIH process had better shape accuracy that specimens formed with the IH process and the FSIF process, especially in Region II and III, as defined in Fig. 2d. In the case of RIH, as discussed in the previous section, local deformations are introduced by both tool rotation and translation. In the case of FSIF, however, because the forming tool traversed along the surface of the workpiece, not only the materials under the forming tool but the materials around are stirred, making the deformation region wider. In addition, the feeding effect also widens the deformation region. As a result, the height of the specimen formed by the FSIF process is lower than expected in each region.

Conclusion

Open-cell nickel foam was successfully formed with the newly proposed method, rotational incremental forming (RIH). The newly proposed process combined the features of incremental hammering process and friction stir welding. In the developed process, the specimen was pressed locally with the forming tool rotating at high speed while the specimen was plastically deformed by the hammering and the rotational motion of the tool. The forming result obtained with the developed process showed significant improvement in forming accuracy compared to the forming result obtained with the conventional incremental hammering.

Acknowledgements The first author thanks Osaka University Scholarship for Super Short Term Study for the financial support.

References

1. Ashby MF, Evans AG, Fleck NA, Gibson LJ, Hutchinson JW, Wadley HNG (2000) Metal foams. Butterworth-Heinemann, Oxford
2. Banhart J (2001) Manufacture, characterisation and application of cellular metals and metal foams. Prog Mater Sci 46(6):559–632
3. Inazawa S, Hosoe A, Majima M, Nitta K (2010) Novel plating technology for metallic foam. SEI Tech Rev 71:23–30
4. Lefebvre LP, Banhart J, Dunand D (2008) Porous metals and metallic foams: Current status and recent developments. Adv Eng Mater 10(9):775–787
5. Liu P, Chen GF (2014) Porous materials. Elsevier, Amsterdam
6. Lobos J, Suzuki S, Nakajima H, Ji Y, Fujii H, Terada D, Tsuji N (2009) Structural change and improvement of the mechanical properties of a lotus-type porous copper by wire-brushing. J Phys: Conf Ser 165
7. Matsubara M, Tanaka S, Nakamura T (1996) Development of incremental sheet metal forming system using elastic tools: principle of forming process and formation of some fundamentally curved shapes. JSME Int J Ser C Dyn Control Robot Des Manuf 39(1):156–163

8. Matsuda K, Koriyama S, Sae-Tae P, Suzuki S, Asakawa M, Ide T, Nakajima H (2013) The effects of structures and mechanical properties of lotus-type porous copper by shot peening process. J Jpn Inst Copp 52(1):92–96
9. Matsumoto R, Tsuruoka H, Otsu M, Utsunomiya H (2015) Fabrication of skin layer on aluminum foam surface by friction stir incremental forming and its mechanical properties. J Mater Process Technol 218:23–31
10. Otsu M, Ichikawa T, Matsuda M, Takashima K (2011) Development of friction stir incremental forming. J Jpn Soc Technol Plast 52(603):490–494. https://doi.org/10.9773/sosei.52.490

Part III
Characterization

Decomposition of Ti and Zr Hydrides Studied by Neutron Diffraction

John Banhart and Clemens Ritter

Abstract TiH_2 and ZrH_2 are the most frequently used blowing agents for foaming low-melting alloys based on aluminium or zinc. In order to tailor the temperature range of decomposition such hydrides are often pre-annealed in Ar or air prior to their use. The decomposition behaviour of such compounds is usually studied by thermogravimetry and calorimetry. We applied neutron diffraction to gather more information about the decomposition of treated and untreated hydrides. In order to avoid strong incoherent scattering effects from hydrogen (^1H), we used deuterated (^2H) variants and could obtain high-quality diffraction patterns. In this way, we were able to quantify the amount of hydrogen for different decomposition stages as well as changes of the lattice structure for a range of Ti- and Zr-based blowing agents.

Keywords Metal foam · Titanium hydride · Zirconium hydride · Neutron diffraction

Introduction

Metal foams can be blown with gas-releasing hydrides that are either added to a melt or admixed to metal powder blends that are compacted and then melted. For low-melting alloys such as aluminium, magnesium, or zinc alloys, titanium or zirconium hydrides have been known for a long time to have a good gas yield in the right temperature range while they are not too expensive [2].

Both hydrides release hydrogen above about 400 °C and exhibit two decomposition peaks, however, ZrH_2 less pronounced than TiH_2 [9, 10]. The first peak is actually not desirable, because gas evolves before most aluminium alloys melt and therefore defects in the foams are created, Fig. 1a. In order to coordinate the melting

J. Banhart (✉)
Helmholtz-Centre Berlin, Hahn-Meitner-Platz 1, 14109 Berlin, Germany
e-mail: banhart@helmholtz-berlin.de

C. Ritter
Institute Laue-Langevin, 71 Avenue Des Martyrs, 38042 Grenoble, France
e-mail: ritter@ill.eu

N. Dukhan (ed.), *Proceedings of the 11th International Conference on Porous Metals and Metallic Foams (MetFoam 2019)*, The Minerals, Metals & Materials Series, https://doi.org/10.1007/978-3-030-42798-6_4

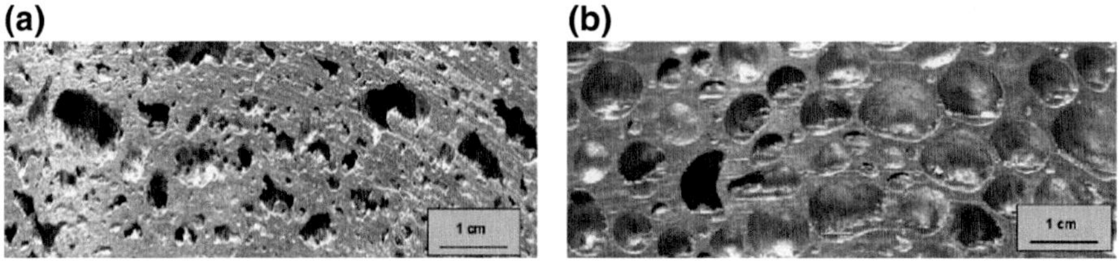

Fig. 1 Aluminium foams after expansion to maximum volume. foamed with, **a**: untreated TiH_2, **b**: TiH_2 treated at 520 °C for 180 min [10]. Treated TiH_2 gives rise to the formation of rounder and more uniform cells

temperature of the alloy to be foamed and the decomposition range of the hydride, an oxidising pre-treatment is useful. Such pre-treatment (or pre-annealing) creates an oxide layer but also releases some hydrogen, which is then not available during the subsequent foaming step of the metal [4].

The decomposition of hydrides has been studied mainly by thermal analysis, X-ray diffraction, TEM, or other methods [3, 6, 7, 8]. The disadvantage of XRD is that hydrogen atoms do not contribute to the information obtained. Neutron diffraction, on the other hand, includes scattering from hydrogen atoms and allows for a structural analysis including the metal lattice and hydrogen. Due to the strong incoherent scattering of neutrons from ^1H, use of ^2H is preferable. Small-angle scattering experiments based on such compounds were successful [1].

In this study, we evaluate the potential of neutron diffraction to clarify the changes during decomposition of hydrides that had been annealed in air or Ar atmospheres. The equilibrium phase diagrams (Fig. 2) show low-temperature α and high-temperature β phases in both Ti-H and Zr-H systems. While the entire diagram is relevant for understanding, for example, hydrogen storage applications, the interest of metal foam application is concentrated on the H-rich side of the diagram, where a cubic δ and a tetragonal ε phase are present.

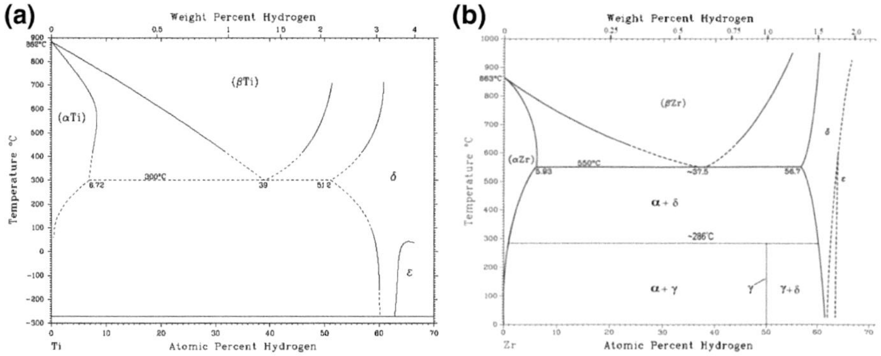

Fig. 2 Phase diagrams of **a** Ti-H [12] and **b** Zr-H [13]

Experimental Procedure

Materials

TiH_2, TiD_2, and ZrD_2 powders were used for the experiments. TiH_2 was purchased from Chemetall, the deuterides were produced by Gesellschaft für Elektrometallurgie (GfE), both Germany, and are the same powders used in a previous study [1]. The powders were characterised in three different states:

- Untreated with a maximum hydrogen/deuterium loading,
- Heat-treated in an Ar atmosphere for various times leading to the removal of H_2 or D_2,
- Heat-treated in air for various times and at various temperatures to allow for oxidation in addition to H_2 or D_2 removal.

Methods

A Netzsch TG209F1 Iris micro-balance was used for thermogravimetric (TG) measurements to determine the hydrogen or deuterium release from the specimens during heating. The mass spectrometer QMS 403 Aëolos attached to the balance allowed us to measure ion currents for $m = 2(H_2)$ and $m = 4(D_2)$. About 50 mg of each hydride was loaded into Al_2O_3 crucibles. A heating ramp of 10 K/min from room temperature up to 900 °C was employed. Prior to each measurement, the specimen chamber was evacuated twice and purged with argon gas. The measurements were performed under an Ar stream of 20 ml/min.

The high-resolution neutron powder diffractometer D1A at ILL was used for the diffraction measurements. The powders were kept in vanadium containers due to largely incoherent (direction independent) scattering of V. Neutrons were monochromatised to 0.139 nm wavelength using Ge crystals, 30-mm high with a take-off angle of 122°. The beam size at the sample site was $20 \times 30\ mm^2$, the neutron flux $2 \times 10^6\ cm^{-2}s^{-1}$.

Diffraction patterns were analysed by Rietveld refinement using software 'FullProf'. In total 16 parameters were varied including the lattice constants and 3 possible occupancies (lattice + tetrahedral and octahedral interstitial positions). Refinements were based on space groups $Fm\overline{3}m$ (#225, fcc + interstitials) or $14/mmm$ (#139, bct + interstitials) as suggested by San Martin [12] and Zuzek [13] for the δ and ε phases, respectively, of both the Ti-H and Zr-H systems. The resulting parameters were the occupancies of all the potential hydrogen sites and the c/a ratio of the metal lattice. We define the relative c/a ratio as $(c/a)_r = c/\left(\sqrt{2}a\right)$.

Results and Discussion

For the deuterides, the peak ion current measured for m = 4(D_2) was about 1000 times larger than for m = 2(H_2) and 45 times larger than for m = 3(DH), indicating that the deuterides were contaminated with hydrogen only on a level of < 2%. Figure 3a shows the ion currents for m = 4, $I(T)$, for three thermal pre-treatment states of TiD_2 (including no treatment) during heating. As reported before, pre-treatment shifts the onset temperature of decomposition to higher temperatures and largely eliminates the first decomposition peak [4]. After reaching 900 °C, most of the gas has been removed. The ion current actually represents the mass change very well with only a slight delay as seen by comparing the directly measured mass $m_d(T)$ given as a percentage of original mass with the relative residual mass $m_i(T)$ calculated from the ion current by

$$m_i(T) = 1 - a\int_0^T I(T')dT', \tag{1}$$

where a was set such that the two curves coincide at 800 °C. Obviously, the mass spectrometer and the balance yield a consistent result and allow for determining the total gas content in the powder that is released up to 900 °C, Fig. 3b. The slight delay of the ion current signal is due to the time the gas needs to pass the spectrometer, while the balance reacts almost instantaneously.

Figure 4 shows diffraction patterns of all three untreated powders. The very high incoherent background for TiH_2 is obvious. The deuterated powders exhibit much less background scattering, which is an advantage when refining the peaks. The peak

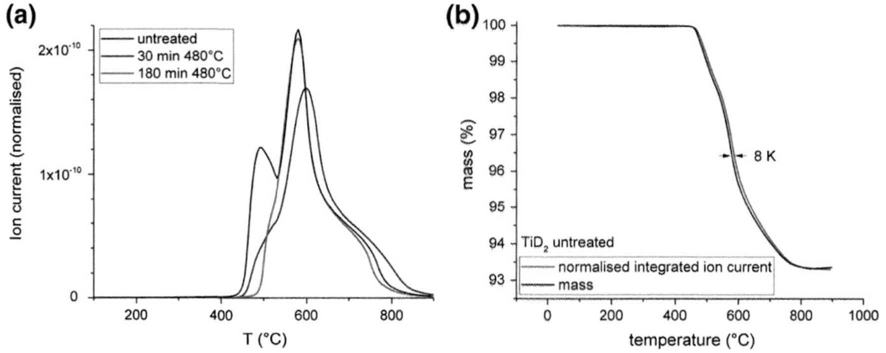

Fig. 3 **a** Ion currents normalised by factor a of Eq. (1) for three differently treated TiD_2 powders (annealed in Ar) during heating at 10 K/min in an Ar atmosphere. **b** mass change during heating at 10 K/min measured with the balance ($m_d(T)$, black) compared to the mass change calculated from the integrated ion current ($m_i(T)$, red). Result is for untreated TiD_2, but treated powders behave in a very similar way

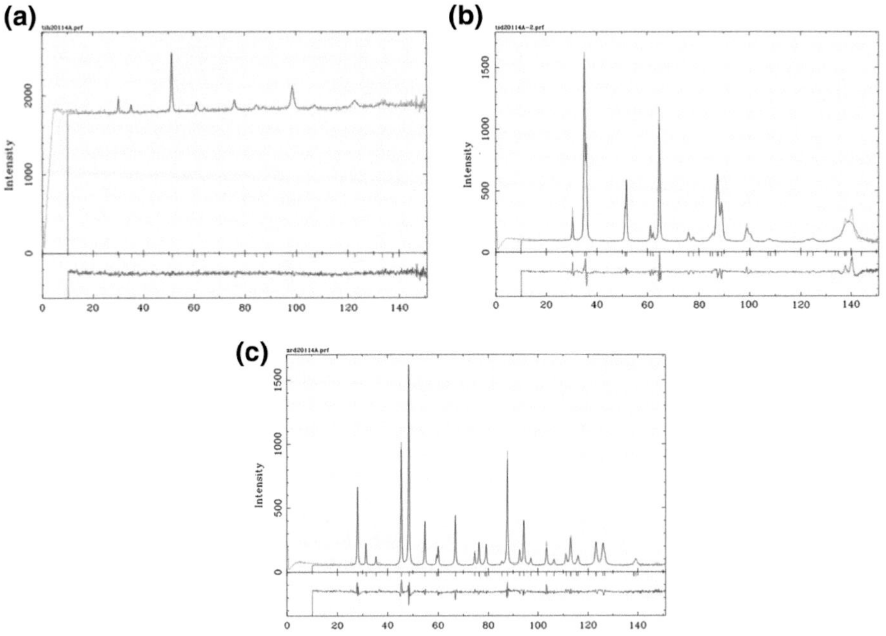

Fig. 4 Diffraction patterns of untreated TiH_2, TiD_2, and ZrD_2 powders (**a**–**c**). The measurement time was 8 h for the hydride, 2 h for the deuterides. Green lines are measured data, red lines Rietveld fits. Residual is shown at the bottom. Angles represent 2Θ

amplitude is 50% higher for the deuterides despite an acquisition time of just ¼ compared to the hydrides.

By doing Rietveld refinements, the H and D contents can be determined quantitatively. Figure 5 shows that for longer pre-treatment time the residual H/D in the powders decreases continuously. We obtain non-zero Rietveld coefficients for the fcc lattice positions and for the tetrahedral lattice sites only. Therefore, as expected, all the hydrogen was found on tetrahedral interstitial lattice sites, none on octahedral sites in any sample. The initial content of the powders also varies: Only ZrD_2 reaches the nominally stoichiometric value of $x = 2$, whereas the Ti-based hydride/deuteride fall short of this value.

The lowest hydrogen contents are found for untreated TiD_2. TGA confirms the neutron diffraction values and actually lie a bit lower ($x = 1.68$). Most likely the low D content is due to insufficient powder charging with deuterium. Some exchange of D by H that might have taken place during storage would also explain the deficiency but is not supported by the TGA measurements for $m = 3$ (suggesting $< 2\%$ H) nor by the low incoherent background in Fig. 4b that would be much higher if significant D had been replaced by H. That TGA yields slightly lower values might be due to some residual D that has not been removed at 900 °C or some oxidation that has led to a slight mass increase due to oxygen traces in the TG chamber.

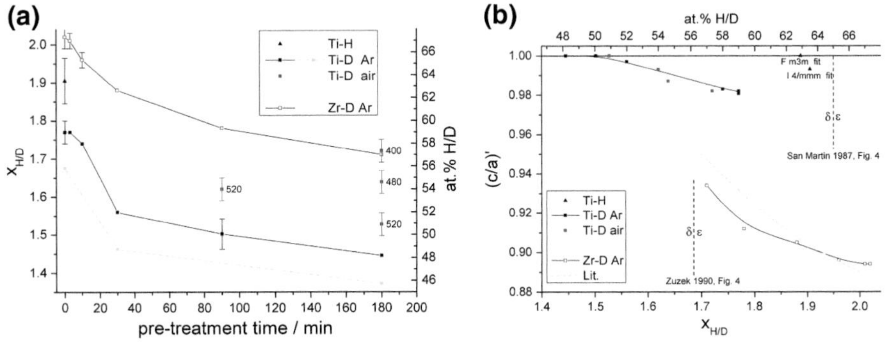

Fig. 5 **a** Hydrogen/deuterium content in various compounds as determined by neutron diffraction and TG (green) for powders treated at 480 °C in an argon atmosphere or in air at the temperature specified. Left axis: stoichiometric content, right axis: at.%. **b** relative c/a ratio displayed as a function of the H or D content

Annealing for 180 min at 480 °C in an Ar atmosphere removes about 11 at.% of the gas in the compound, which is just 1/6 of the total content. This explains why annealing does not influence foam expansion notably [11]. Annealing in air instead of argon (what is actually done in industrial practice) leads to even lower gas losses. For example, 180 min annealing at 480 °C limits the gas loss during pre-treatment in air to just above 4 at.%. Higher temperatures or longer times increase gas losses. The effect of air has been explained by the formation of an oxide shell on each hydride particle [5].

TiH_2 was not systematically studied by neutron diffraction due to the large incoherent scattering background. The only measurement on the untreated powder (Fig. 3a) yields a composition of $TiH_{1.9}$ with a rather high uncertainty. ZrD_2 exhibits the highest D content of all the samples studied. Pre-treatment in Ar atmosphere reduces the content of deuterium dissolved in the matrix by approximately the same percentage as in TiD_2.

Rietveld refinements were carried out based on a cubic δ and a tetragonal ε structure as the phase diagrams (Fig. 2) suggest. The resulting $(c/a)_r$ ratio is given in Fig. 5b. The closer the value is to 1 the less do the two fits differ and allow us to determine whether a δ or a ε structure is present. The lowest values are obtained for ZrD_2. The phase diagram (Fig. 2b) actually predicts that for $x_D > 1.7$ the lattice is tetragonal, which is confirmed by our measurements. The broken line in Fig. 5b shows values taken from Zuzek [13] and confirms the current results. For TiD_2, the situation is different. According to San Martin [12], the lattices should be cubic for the low $x_D < 1.8$ observed but the neutron diffraction signal seems to show a trace of tetragonal distortion. Fits of the TiH_2 data in Fig. 5b carried out in both the cubic and tetragonal variants yield the same fitting residual and show that for near-cubic structures the distinction can be difficult. One might suspect such uncertainty for TiD_2 too, but the $(c/a)_r$ values are smaller and do vary systematically. Moreover, the oxidised samples show the same values. A possible explanation is that some of the larger powder particles contain more D than others and contribute to a slight tetragonal signature in the diffraction patterns.

Summary

Diffraction experiments were used to understand the modification of hydrides used for metal foaming during pre-treatments. The hydrogen/deuterium contents measured roughly agree with those measured by thermogravimetry and mass spectrometry. Pre-annealing powders in air leads to less hydrogen losses than annealing in an Ar atmosphere due to the oxide layer created on each particle. The hydrogen lost during pre-annealing accounts for less than 10% of the total hydrogen, in industrially treated powders rather ~4%. Thus, this loss can be accepted in view of the known benefits of pre-annealing.

Acknowledgements Biljana Matijasevic, Tillmann Neu, and Nadine von der Eltz prepared the powder samples. Claudia Leistner performed the TG.

References

1. Banhart J, Bellmann D, Clemens H (2001) Investigation of metal foam formation by microscopy and ultra small-angle neutron scattering. Acta Mater 49(17):3409–3420
2. Banhart J (2013) Light-metal foams—history of innovation and technological challenges. Adv Eng Mater 15(3):82–111
3. Gambini M, Montanari R, Richetta M, Stilo T, Varone A, Vellini M (2018) Hydrogen release from oxidized titanium hydride. Mater Sci Forum 941:2203–2208
4. Gergely V, Clyne TW (2000) The FORMGRIP process: foaming of reinforced metals by gas release in precursors. Adv Eng Mater 2(4):175–178
5. Jiménez C, García-Moreno F, Pfretzschner B, Klaus M, Wollgarten M, Zizak I, Schumacher Tovar M, Banhart J (2011) Decomposition of TiH_2 studied in situ by synchrotron X-ray and neutron diffraction. Acta Mater 59(16):6318–6330
6. Kaciulis S, Soltani P, Mezzi A, Montanari R, Lapi G, Richetta M, Varone A, Barbieri G (2017) Oxidative treatment effect of TiH_2 powders. Surf Interface Anal 50:1195–1199
7. Ma M, Wang L, Tang B, Lyu P, Xiang W, Wang Y, Tan X (2018) Kinetics of hydrogen desorption from titanium hydride under isothermal conditions. Int J Hydrogen Energy 43:1577–1586
8. Ma M, Wang L, Wang Y, Xiang W, Tang B, Lyu P, Tan X (2018) Oxidation behaviour of zirconium hydride and its influence on the thermal desorption kinetics. Corros Sci 134:199–210
9. Matijasevic B, Görke O, Schubert H, Banhart J (2005) Zirconium hydride—a possible blowing agent for making aluminium alloy foams. In: Nakajima H, Kanetake N (eds) Porous metals and metal foaming technology. The Japan Institute of Metals, pp 107–110
10. Matijasevic B, Banhart J, Fiechter S, Görke O, Wanderka, N (2006) Modification of titanium hydride for improved aluminium foam manufacture. Acta Materialia 54(7):1887–1900

11. Matijasevic B, Banhart J (2006) Improvement of aluminium foam technology by tailoring of blowing agent. Scripta Materialia 54(4):503–508
12. San Martin A, Manchester FD (1987) The H-Ti (Hydrogen-Titanium) system. Bull Alloy Phase Diagr 8(1):30–42
13. Zuzek E, Abriata JR, San Martin A, Manchester FD (1990) The H-Zr (Hydrogen-Zirconium) system. Bull Alloy Phase Diagr 11(4):385–395

Microstructural Effects on Compressive Behavior and Deformation Band Propagation in Open-Pore Metal Foams

Alexander Martin Matz, Bettina Stefanie Matz and Norbert Jost

Abstract The mechanical properties of investment casted open-pore metal foams are studied on the example of the binary alloy Al–11Zn. The samples are subjected to two different heat treatments, resulting in a state of high strength and low ductility and a state of lower strength and high ductility. Their effects are investigated by compression testing and are correlated to their microstructure. As recently reported in literature, there exist characteristic differences in the stress–strain behavior. This issue has been researched in the course of the present work in detail. We found that it can be directly related to the formation and propagation of deformation bands, and herein locally occurring plastic deformation events. In particular, we observe strength and ductility of the base material in combination with the morphological structure as the dominating factors for the evolution of deformation in metal foams.

Keywords Metallic foams · Aluminum · Zinc · Heat treatment · Aging · Mechanical properties · Deformation band

Introduction

Metal foams are attractive engineering materials, which combine low weight with a good mechanical strength [1–4]. They have been one of the research highlights in materials engineering ever since the late 80s of the last century [3–9]. The present work is concerned with open-pore metal foams. They consist of a rigid strut network, which defines an interconnected population of cells. The potential of open-pore metal

A. M. Matz (✉) · B. S. Matz · N. Jost
Institute for Materials and Material Technologies, Pforzheim University of Applied Sciences, Tiefenbronner Straße 65, Pforzheim 75175, Germany
e-mail: alexander.matz@hs-pforzheim.de

B. S. Matz
e-mail: bettina.matz@hs-pforzheim.de

N. Jost
e-mail: norbert.jost@hs-pforzheim.de

N. Dukhan (ed.), *Proceedings of the 11th International Conference on Porous Metals and Metallic Foams (MetFoam 2019)*, The Minerals, Metals & Materials Series, https://doi.org/10.1007/978-3-030-42798-6_5

foams in a wide variety of applications has been emphasized in the literature [10–14]. Their mechanical properties are always a determining factor for choosing them as a constructional element. We have recently described how one can make these materials using the model alloy Al–11Zn [15]. We have shown how irreversible plastic strain accumulation occurs during compression testing [16]. We have also demonstrated how the local Al–11Zn microstructure, which evolves during post-processing heat treatments, affects the macroscopic mechanical behavior of an open-pore metal foam during compression testing [17]. The microstructure mainly defines the deformation behavior of bulk material. In foams, it is additionally affected by the structural architecture of the cellular material leading to their capability to effectively dissipate energy under compressive loading. The stress–strain curves of metal foams under compressive loading show the three typical stages of formation:

(1) elastic–plastic region,
(2) plateau region, and
(3) densification region,

as commonly described in literature [2, 18–20]. During compressive loading of the specimen, we preferentially observe a localization of plastic strain in single struts. They underlie such a pronounced deformation that single foam cells collapse and one or a few deformation bands form throughout the entire foam [21]. The stress–strain curve in most cases indicates the initial cell collapse [4] (the corresponding stress is referred to as plastic collapse stress σ_{PC}). It shows a local maximum (which separates the elastic–plastic and the plateau region) due to the sudden stress drop caused by a failing strut. This is characteristic of metal foams subjected to compressive loading, and it is higher the more brittle the base material, such as the aluminum alloy AA6101 in aged conditions [21, 22]. Ductile materials, such as Al in a chemical purity of 99.99% or Al–12Si [23–25], rather show a smooth transition of the elastic–plastic to the plateau regime of the stress–strain curve. Besides the base materials, the thermal treatment seems to be another parameter affecting this behavior [17, 26].

From what has been outlined so far, it is clear that the microstructure and the structural architecture of the cellular material are the factors dominating the fashion of the stress–strain curves. The stress–strain curves are, however, just what we gain from the test procedure; and this is actually a result of the deformation process taking place within the specimen in general. For a profound understanding of the underlying correlations, we manufacture open-pore Al–11Zn foams and apply two different heat treatments. Furthermore, we investigate the evolution of deformation in the foams during compressive loading and observe significant effects of the microstructure on the deformation band propagation during compressive loading.

Experimental Procedure

The material used was the binary aluminum alloy Al–11Zn. It was alloyed by HMW Hauner Metallische Werkstoffe from Al and Zn, each with a nominal purity of 99.99%. The materials are the same as those used by Matz et al.; for details of the chemical composition, refer to [16].

Open-pore Al–11Zn foams were fabricated through a modified investment casting process. Description, that is more detailed, can be found in [15]. Casting was carried out by a centrifugal casting machine of the type Vacutherm-3,3-Titan from Linn High Therm GmbH using a crucible of the type DKV-2 LT. Subsequent to casting, the molds containing the samples were water-quenched. Mold removal and cleaning of the samples were carried out in a water jet machine of the type HW 70 from KWS Kächele GmbH, and afterward by a chemical treatment in tetranatriumethylendiamintetraacetate ($C_{10}H_{12}N_2Na_4O_8$). Samples were cut in dimensions of $30 \times 30 \times 20$ mm^3 by wire electro-discharge machining [27]. A homogenization treatment at a temperature of $T = 400$ °C lasting $t = 120$ min in connection with water-quenching was applied to all samples. Half of them were aged at room temperature (samples *A(RT)*) to achieve finely dispersed GP zones [28]. The rest pass through an additional annealing treatment at a temperature of $T = 150$ °C for $t =$ 120 min (samples *A (150/120)*) in order to obtain a microstructure with partially dissolved GP zones. Details on the effects of heat treatments on the microstructure are described elsewhere [17]. The pore density of all samples was $\rho_P = 10$ ppi, and the relative density was in a range of $7.5\% \leq \rho_{rel} \leq 11.0\%$.

Uniaxial compression tests were performed with a TesT universal test rig P112.20 KN.H at room temperature and ambient air. The displacement rate was set to be 0.9 mm/min, which corresponds to an initial strain rate of 5×10^{-4} s^{-1}. Precision-grinded steel plates (coated by a graphite lubricant) were used to apply compression forces to the samples (Fig. 1). The testing procedure was documented with a camera

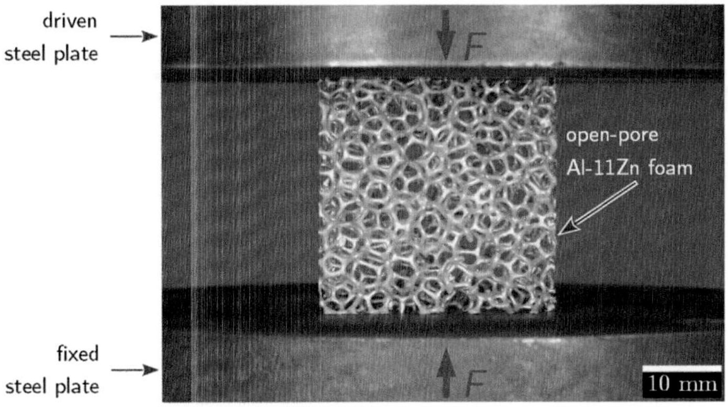

Fig. 1 Experimental setup for compression tests of open-pore Al–11Zn foams, which corresponds to the setup described in Ref. [16]

of the type Cyber-SHOT DSC-HX50V from Sony, Corp. in order to evaluate the evolution of deformation within the whole foam specimen. Furthermore, close-up views of relevant parts of the deformed samples were taken with a digital microscope (VHX 500FD from Keyence NV/SA).

Results

Compressive Properties

The stress–strain curves (Fig. 2) of the foams under compressive loading show in general the three abovementioned typical stages of formation. At the beginning of deformation, the samples aged at room temperature have a linear increase in stress. Thereafter, we notice an increasing portion of plasticity, which is a consequence of strain hardening [16], until the tangent modulus turns zero, when reaching the plastic collapse strain σ_{PC}. The plateau region follows up, when a sudden stress drop occurs and subsequently continues as an oscillating curve. At higher strains, the distinct rise in stress can be identified as densification region. The other samples, which passed through an additional annealing treatment at a temperature of $T = 150$ °C, exhibit a different curve fashion with the characteristic smooth transition from the elastic–plastic region into the plateau region. The strain at which the transition occurs is hard to judge, but it is in an order of $\varepsilon \approx 10\%$ as also reported previously for ductile Al foams [24]. In contrast to the samples aged at room temperature, no oscillating fashion of the curve can be detected in the plateau region. The densification of the foams is mainly a function of relative density [17, 18] and is, hence, in the same order as the samples *A(RT)*.

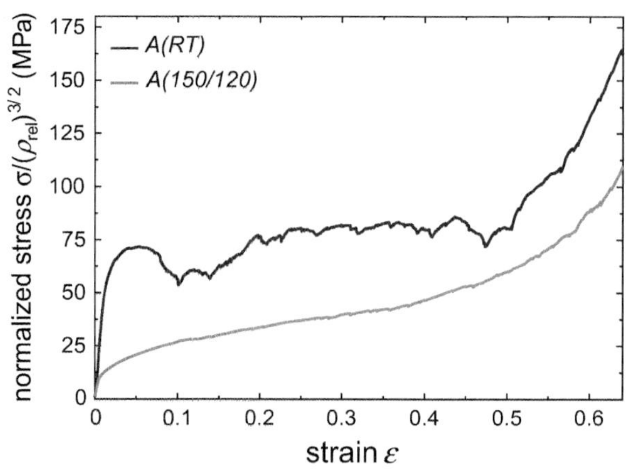

Fig. 2 Compressive stress–strain behavior of open-pore Al–11Zn foams in heat treatment conditions *A(RT)* and *A(150/120)*

Local Deformation Behavior

Compressive loading of open-pore Al–11Zn foams leads to a preferably local formation of deformation bands. Within the concerned specimen areas, single struts are deformed to such a high extent until a fatal collapse of the corresponding cell occurs and/or until the locally reached strain hardening leads to spreading of the deformation band into prior non-effected specimen areas. Figure 3a–c shows how the deformation in an open-pore Al–11Zn foam aged at room temperature evolves on the example of different deformation stages. The corresponding stress–strain diagram is shown in Fig. 3g. The deformation bands, which are active in the different deformation stages, and the hitherto initiated fractures are labeled in the images. The sample displayed in Fig. 3a resamples the reference state at a strain of $\varepsilon = 0$ immediately before the compression test started. When applying compressive loading, at first, a homogenous deformation within the whole specimen occurs, which still seems to be accompanied by a portion of elastic strain ($\varepsilon \leq 1.37\%$). Thereafter, a first deformation band forms and arranges in the upper part of the sample oriented in an angle of $\alpha \approx 20°$ across the direction of force application. A further deformation band forms when reaching a strain of $\varepsilon \leq 3.55\%$ oriented in an angle of $\alpha \approx 15°$ across the direction of force application. It is located in the lower part of the sample and it is connected near the left sample surface with the first deformation band formed (this state is shown in Fig. 3b at a strain of $\varepsilon = 6.50\%$). The deformation itself, however, is still restricted to the deformation band, which is located in the upper part of the specimen. Within this band and in the direct adjoining parts, fractures of the struts occur, which are labeled by arrows in Fig. 3b. At a nominal strain of $\varepsilon = 5.32\%$, the ongoing deformation of the sample is of the same order in both active deformation bands. Not before reaching $\varepsilon = 8.22\%$, the main deformation shifts to the deformation band, which is located in the lower part of the specimen. This deformation band is distinctively silhouetted against the sample parts above and below. Furthermore, fracture initiation and fracture occur within the deformation bands and at adjacent regions where the strut orientation itself leads to a highly rigid structure, and hence restrains plastic deformation (this state is shown in Fig. 3c at a strain of $\varepsilon = 11.04\%$). Afterward, entire cell planes, containing the afore fractured struts, collapse and the deformation band spreads into regions of the sample at which no prior macroscopic deformation could be observed. In direct comparison to the sample in cold-aged condition, the progressive deformation behavior of the sample in the condition with partially dissolved GP zones is shown in Fig. 3d–f. The corresponding stress–strain diagram is displayed in Fig. 3h. Figure 3d shows the initial state at a strain of $\varepsilon = 0$. When compression testing started, in the first instance (as well as observed in the cold-aged condition), the sample deforms homogeneously. Subsequently and as a gradual process, the deformation events get concentrated in the middle of the sample. Correspondingly, at parts of the sample, which are in direct contact with the compression plates, minor deformation occurs, and when reaching a strain of $\varepsilon = 5.79\%$ no deformation events can be observed anymore. Thereupon, an increasing localization in the form of a distinctly wide shaped deformation band

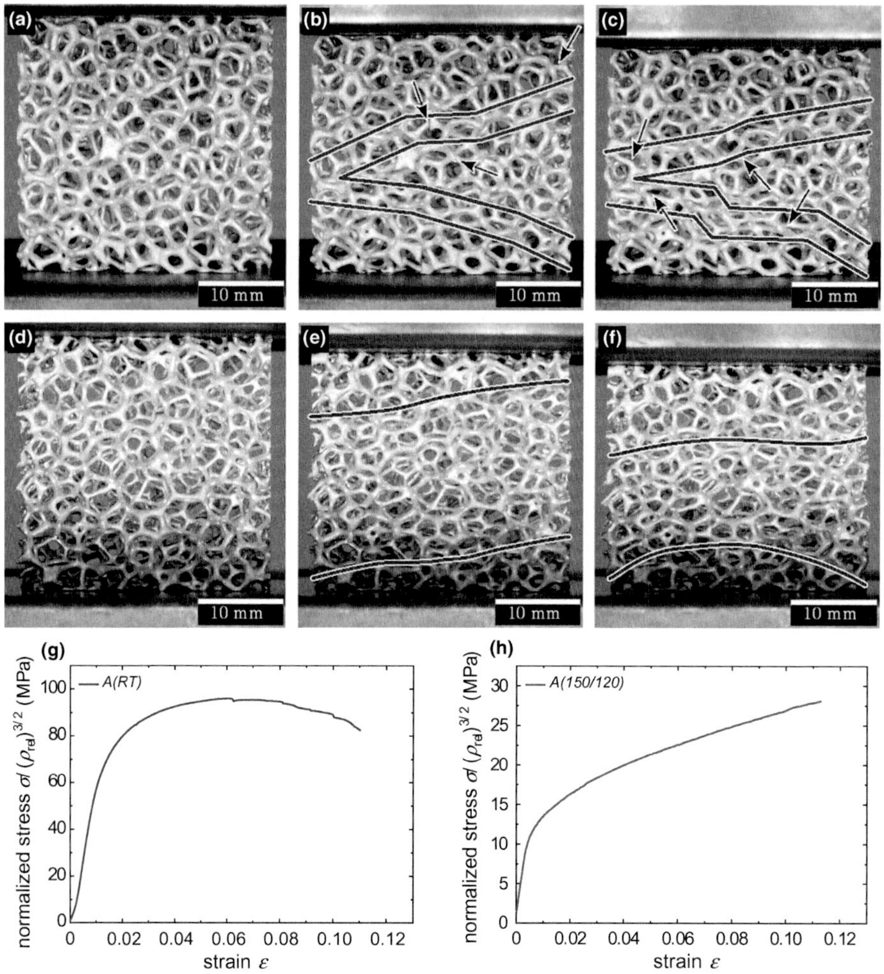

Fig. 3 Formation and propagation of deformation bands, and fracture initiation during compression testing of open-pore Al–11Zn foams: **a** reference state ($\varepsilon = 0$), **b** compressed to $\varepsilon = 6.50\%$, **c** compressed to $\varepsilon = 11.04\%$, and **g** corresponding stress–strain behavior of sample *A(RT)*, as well as **d** reference state ($\varepsilon = 0$), **e** compressed to $\varepsilon = 6.32\%$, **f** compressed to $\varepsilon = 11.29\%$, and **h** corresponding stress–strain behavior of sample *A(150/120)* (deformation bands are marked by lines and fractures are marked by arrows)

can be noticed, which stretches across about ¾ of the sample height (this state is shown in Fig. 3e at a strain of $\varepsilon = 6.32\%$). At $\varepsilon \approx 7.39\%$, the deformation band gradually relocates into the lower third of the sample. The major deformation events proceed in the center of the deformation band. When reaching a strain of $\varepsilon \approx 8.37\%$, the deformation stagnates in the lower part of the sample and the deformation band starts shifting upward. Ongoing strain intensifies this behavior as can be seen in Fig. 3f ($\varepsilon \approx 11.29\%$) and leads to a pronounced bending of single struts until they finally touch each other. Not before a compaction of the affected cell planes occur, the deformation band relocates into sample parts, which are macroscopically much less deformed.

The deformation events in both samples differ totally as pointed up in Fig. 4. The cold-aged sample is considerably more brittle. It shows a pronounced sensitivity to fracture events as illustrated in Fig. 4b (this is a detail view of Fig. 4a, which corresponds to the final state of compression of the cold-aged sample, as shown in Fig. 3c). The highlighted fracture *I* is the result of a shearing dominated strain, whereas fractures *II* and *III* are the consequence of bending-dominated strain. The sample in the condition with partially dissolved GP zones is considerably ductile in contrast to the cold-aged sample. Prior to the collapse of the first cell plane, no fractures can be detected. We rather observe a pronounced local plastic deformation, as shown in the marked region of the sample in Fig. 4c. A detail view of the relevant strut (which

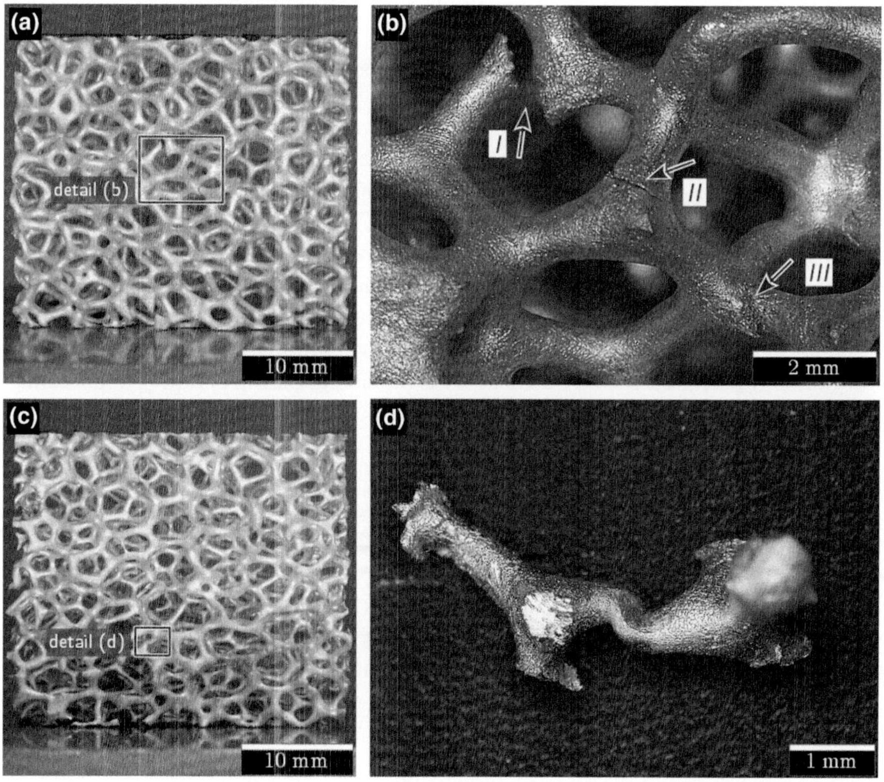

Fig. 4 Deformation phenomena of struts within a deformation band of open-pore Al–11Zn foams: **a** sample *A(RT)* compressed to $\varepsilon = 11.04\%$ (cf. Fig. 3c), **b** close-up view of labeled sample area in (**a**), **c** sample *A(150/120)* compressed to $\varepsilon = 11.29\%$ (cf. Fig. 3f), and **d** close-up view of labeled sample area in (**c**)

was extracted for better presentation) is displayed in Fig. 4d. The alteration of shape is clearly visible and likewise a result of shearing dominated strain.

Discussion

The deformation of open-pore metal foams relies on the formation of local deformation bands. Regarding both microstructures considered in the present study, we clearly observe differences in shape and in propagation of the deformation bands during compressive loading. At the beginning, the deformation band in the specimen *A (150/120)* nearly involves the whole sample. It localizes or concentrates gradually as a function of strain ε (cf. Fig. 3e, f). This behavior also becomes noticeable in the stress–strain diagram (cf. Fig. 2) at which a continuous increase of stress σ occurs in the plateau region. It is a consequence of the strain hardening of the base material. It seems that the sample areas, which are next to the compression plates, firstly underlie a strain hardening which spreads into the center of the specimen as a function of strain ε. Against this background, we notice a strain hardening rate $k_{\mathrm{Pl}} = d\sigma / d\varepsilon$ in the plateau region of $k_{\mathrm{Pl}} = 1.7\,\mathrm{MPa}$ [17]. In contrast, the strain hardening rate in the plateau region is only about two-thirds of that in the specimen *A(RT)* [17]. Its deformation bands (one or more) are formed in a dimension of approximately one cell plane in the specimen *A(RT)* (cf. Fig. 3b, c). The deformation band concentrates locally right from the outset, as numerously reported in literature for metal foams with open and closed cells [20, 21, 29–32]. Strain hardening takes place within the affected specimen area until single struts fracture. This leads to a local structural softening of the foam and in consequence to an oscillating behavior of the stress–strain curve (cf. Fig. 2). Based on this, we conclude that during compression loading, there is a prior base material controlled strain hardening in the specimen *A (150/120)*, whereas there is an interrelated strain hardening of base material and softening of foam structure in the specimen *A(RT)*.

The deformation bands are in general oriented in an angle of $15^\circ \leq \alpha \leq 20^\circ$ transverse to the direction of applied compressive force as also reported previously [29, 31, 33]. In analogy to slip planes in crystals, we interpret this phenomenon to be related to the packing of the polydisperse foam's polyhedral cells in space but here on the mesostructural scale. The deformation itself is concentrated on the weakest part of the foam and evolves with increasing nominal strain ε by

(1) deformation of single struts to such an extent that a collapse of the belonging cell occurs,
(2) a pronounced local strain hardening of the bulk material, and/or
(3) a local hardening of the foam structure itself, which is related to a geometrical beneficial arrangement of partially deformed struts being more stable than adjacent specimen parts.

The aspects (1) and (3) are the major active mechanisms for the formation and propagation of deformation bands in samples *A(RT)*, whereas (2) and (3) are predominant in the samples *A(150/120)*. This explains on the one hand the differing shapes of the deformation bands (Fig. 3) and on the other hand their characteristic stress–strain curve behaviors, such as a noticeable plastic collapse in *A(RT)* samples or a noticeable linear strain hardening rate in the plateau region in *A(150/120)* samples.

Conclusion

In the present study, we investigate how the microstructure of Al–11Zn affects the compressive behavior and deformation band propagation in open-pore metal foams. From the results obtained in this work, the following main conclusions can be drawn:

(1) The major deformation events concentrate on one or a few parts of the sample (deformation bands) until either a collapse of the cell plane occurs resulting in a structural softening of the foam or the foam's base material undergoes a strain hardening to such an extent that the deformation migrates to an adjacent weaker part and further deformation band propagation gets activated;
(2) In a metal foam with brittle base material, the deformation locally remains until fracture of struts and/or collapse of cells or cell planes occur—its deformation is dominated by structural softening;
(3) In a metal foam with ductile base material, the deformation is rather homogenous within the whole sample and gradually localizes in a deformation band as a function of strain—its deformation is dominated by strain hardening of its base material.

Acknowledgements Funding through European Regional Development Fund (ERDF) and the Ministerium für Wissenschaft, Forschung und Kunst Baden-Württemberg within the research center ZAFH InSeL and the program Mittelbau at HAW is gratefully acknowledged. The authors thank Prof. K. Oßwald and D. Kammerer of Pforzheim University of Applied Sciences for machining the samples. They are also thankful for material contribution of Foampartner Reisgies Kunststoffe GmbH and DuPont Polymer Powders Switzerland Sàrl.

Conflict of interest The authors declare that there is no conflict of interest regarding the publication of this paper.

References

1. Ashby MF (1983) The mechanical properties of cellular solids. Metall Trans A 14(9):1755–1769
2. Banhart J, Baumeister J (1998) Deformation characteristics of metal foams. J Mater Sci 33(6):1431–1440
3. Banhart J (2013) Light-metal foams—history of innovation and technological challenges. Adv Mater Eng 15(3):82–111

4. Gibson LJ (2000) Mechanical behavior of metallic foams. Annu Rev Mater Sci 30(1):191–227
5. Ashby MF, Evans A, Fleck NA, Gibson LJ, Hutchinson JW, Wadley HNG (2000) Metal foams—a design guide. Butterworth-Heinemann, Woburn
6. Banhart J (2001) Manufacture, characterisation and application of cellular metals and metal foams. Prog Mater Sci 46(6):559–632
7. Gibson LJ, Ashby MF (2001) Cellular solids: structures and properties. Cambridge University Press, Cambridge
8. Lefebvre L-P, Banhart J, Dunand DC (2008) Porous metals and metallic foams: current status and recent developments. Adv Mater Sci 10(9):775–787
9. Dukhan N (ed) (2013) Metal foams: fundamentals and applications. DEStech Publications, Lancaster
10. Han X-H, Wang Q, Park Y-G, T'joen C, Sommers A, Jacobi A (2012) A review of metal foam and metal matrix composites for heat exchangers and heat sinks. Heat Transf Eng 33(12):1–20
11. Zhao CY (2012) Review on thermal transport in high porosity cellular metal foams with open cells. J. Heat Mass Transfer 55(13–14):3618–3632
12. Betts C (2012) Benefits of metal foams and developments in modelling techniques to assess their materials behaviour: a review. Mater Sci Technol 28(2):129–143
13. Müller DW, Matz AM, Jost N (2013) Casting open porous Ti foam suitable for medical applications. Bioinsp Biomim Nanobiomater 2(2):76–83
14. Yang G-F, Song K-Y, Joo S-K (2015) Ultra-thick Li-ion battery electrodes using different cell size of metal foam current collectors. RSC Adv 5:16702–16706
15. Matz AM, Mocker BS, Müller DW, Jost N, Eggeler G (2014) Mesostructural design and manufacturing of open-pore metal foams by investment casting. Adv Mater Sci Eng 2014(421729):1–9
16. Matz AM, Matz BS, Jost N, Eggeler G (2018) On the accumulation of irreversible plastic strain during compression loading of open-pore metallic foams. Mater Sci Eng A 728:40–44
17. Matz AM, Matz BS, Parsa AB, Jost N, Eggeler G (2019) On the effects of microstructure on the mechanical properties of open.pore Al-11Zn foams. Mater Sci Eng A 759:552–564
18. Matz AM (2017) Werkstoff-und morphologieabhängige Einflüsse auf die mechanischen Eigenschaften von offenporigen Al-11Zn-Schäumen. Shaker, Herzogenrath
19. Miller RE (2000) A continuum plasticity model for the constitutive and indentation behaviour of foamed metals. Int J Mech Sci 42(4):729–754
20. Schüler P, Fischer SF, Bühring-Polaczek A, Fleck C (2013) Deformation and failure behaviour of open cell Al foams under quasistatic and impact loading. Mater Sci Eng A 587:250–261
21. Jang WY, Kyriakides S (2009) On the crushing of aluminum open-cell foams: Part I experiments. Int J Solids Struct 46(3):617–634
22. Zhou J, Shrotriya P, Soboyejo WO (2004) Mechanisms and mechanics of compressive deformation in open-cell Al foams. Mech Mater 36(8):781–797
23. Despois J-F, Mueller R, Mortensen A (2006) Uniaxial deformation of microcellular metals. Acta Mater 54(16):4129–4142
24. San Marchi C, Mortensen A (2001) Deformation of open-cell aluminum foam. Acta Mater 49(19):3959–3969
25. San Marchi C, Despois J-F, Mortensen A (2004) Uniaxial deformation of open-cell aluminum foam: the role of internal damage. Acta Mater 52(10):2895–2902
26. Zhou J, Gao Z, Cuitino AM, Soboyejo WO (2004) Effects of heat treatment on the compressive deformation behavior of open cell aluminum foams. Mater Sci Eng A 386(1–2):118–128
27. Matz AM, Kammerer D, Jost N, Oßwald K (2016) Machining of metal foams with varying mesostructure using Wire EDM. Proc CIRP 42:263–267
28. Garwood RD, Davies AL (1960) The effect of structural imperfections on the age-hardening of aluminium-zinc alloys. J Inst Met 88(7):311–317
29. Bart-Smith H, Bastawros AF, Mumm DR, Evans AG, Sypeck DJ, Wadley HNG (1998) Compressive deformation and yielding mechanisms in cellular Al alloys determined using X-ray tomography and surface strain mapping. Acta Mater 46(10):3583–3592

30. Bastawros A-F, Bart-Smith H, Evans AG (2000) Experimental analysis of deformation mechanisms in a closed-cell aluminum alloy foam. J Mech Phys Solids 48(2):301–322
31. Mu Y, Yao G, Liang L, Luo H, Zu G (2010) Deformation mechanisms of closed-cell aluminum foam in compression. Scripta Mater 63(6):629–632
32. Mukherjee M, Kolluri M, Garcia-Moreno F, Banhart J, Ramamurty U (2009) Strain hardening during constrained deformation of metal foams—effect of shear displacement. Scripta Mater 61(7):752–755
33. Kolluri M, Karthikeyan S, Ramamurty U (2007) Effect of lateral constraint on the mechanical properties of a closed-cell al foam: I experiments. Metall Mater Trans A 38(9):2006–2013

Mechanical Characterization of Hyper-cubic Models Created with Direct Metal Laser Sintering Method

Jeongho Choi

Abstract The objective of this study was to determine mechanical properties of two models of hypercubes defined as Type 1 (core-filled) and Type 2 (core-spaced) made with direct metal laser sintering (DMLS) method. To manufacture these models, aluminum alloy AlSi10Mg powder was used. Each model was created by adding consecutive layers vertically with DMLS. After quasi-static compression, Type 1 model had 19% higher elastic modulus, 12% higher compressive yield strength, and 51.6% greater elongation than Type 2 model. From our experiments, we found that the following two factors could weaken these models. First, metals melted by DMLS were not connected to each other securely. Second, anisotropy in specimens created by DMLS was a significant factor that affected their stiffness and strength. These two models could be utilized in a sandwich core structure. Further studies are needed to investigate their bending and shear properties more deeply. Upgraded 3D printing technologies in the future might help us make more precise, complex structures.

Keywords Cellular solids · Sandwich core · Unit cell · Hypercube · 3D printing

Introduction

Generally, solids have a stereoscopic structure. The stereoscopic structure is a model based on two-, three-, and four-dimensions in space. Two-dimensions has two axes such as a horizontal and vertical directions in a plane. Three-dimension has three axes such as x, y, and z directions. It can be presented by a vector. Four-dimension has four axes that can be thought as certain optical illusions. For example, we can take a cubic box and then stretch and fold every corner to make a two-dimension plane. This folding-up creates another cubic box. This shape is called a tesseract, hypercube, or 8-cell regular octachoron. These shapes are formed based on certain geometrical concept in mathematics. These shapes can be used to make a unit cell model such as

J. Choi (✉)
School of Mechanical Engineering, Kyungnam University, 7 Kyungnamdaehak-ro, Masanhappo-gu, Changwon-si, Gyeongsangnam-do 51767, Republic of Korea
e-mail: choicaf@kyungnam.ac.kr

N. Dukhan (ed.), *Proceedings of the 11th International Conference on Porous Metals and Metallic Foams (MetFoam 2019)*, The Minerals, Metals & Materials Series, https://doi.org/10.1007/978-3-030-42798-6_6

a honeycomb or an open-cell with a sandwich core structure. Hypercubes originate from a point of a hypercube with dimension zero [1, 2]. If this point is stretched to another point, it makes a line which is a unit hypercube of dimension one. If the line stretches out in a perpendicular direction, it makes a square which is a unit hypercube of dimension two. If the square is stretched in a perpendicular direction, it makes a cube which is a unit hypercube of dimension three. When the cube is stretched to a fourth dimension, it is called a four-dimensional hypercube or a unit tesseract. In the field of material engineering, one of the simplest lightweight truss models is a truss cubic which has been defined by Gibson and Ashby [3]. It is a model of a hexagonal truss. It is an ideal solution for honeycomb, open cell, and closed cell structures. This model is based on a 3-dimensional stereoscopic structure. From the truss model, we can create a hypercube truss model with a hexagonal truss outside and a hexagonal truss inside. Thus, this paper focuses on this hypercube concept to create a unit model to apply in a sandwich core structure. If a tesseract is composed of two regular hexahedrons (one forms the outer structure while the other forms the inner structure) with or without diagonal truss, then we can define two types of model as core-filled and core-spaced shapes. That is, the type of model depends on whether or not there is a truss in the diagonal direction. These two models are defined as type 1 and type 2. According to recent papers about mechanical properties of 3D printed structures made with powders, there are no particular advantages or disadvantages for either shape. Instead, researchers are trying to determine how to make the shape easily, save time, create a complex shape without any limitations, and produce high-quality parts for use in aerospace or biomedical industries. They are looking to avoid high costs, having limited space to make the product, using high cost materials, such as metal powders, buying more high quality equipment to produce a high quality product, having to hire professional engineers to control the equipment, etc. 3D printing is not some kind of magic that can create anything. Skill and thought are required when selecting techniques and materials. The most important factor that needs to consider is what the application is. Depending on producers' skill and materials, the quality of the product can be different.

Recently, 3D-printing has been announced as a revolution in manufacturing. It has been advanced more and more. There are many widely used 3D-printing techniques, such as Fused Deposition Modeling (FDM) [4], Selective Laser Sintering (SLS) [5], Direct Light Processing (DLP) [6], StereoLitohgraphy (SLA) [6], Laminated Object Manufacturing (LOM) [6, 7], StereoLithography (SL) [7], Mask Projection StereoLithography (MPSL) [7], 3-Dimensional Printing (3DP) [7], Droplet Deposition Manufacturing (DDM) [7], and Fused Filament Fabrication (FFF) [8]. Based on these techniques, new skills are developed and announced continuously. These days, Direct Metal Laser Sintering (DMLS) is a common skill used in 3D printing is [9]. It uses lasers with metal powders. These lasers can fuse metal powders into drips of liquid. The drip metal is added layer by layer to make a shape.

To validate a product's mechanical properties, it is required to compare its tension and compression tendencies with a specimen made by the ASTM standard. Many researchers have found that specimens made by 3D printing have an anisotropy with inconsistent mechanical properties. This occurs even after using different materials and 3D printing techniques. A product's final properties depend greatly on the technician's skill, equipment, materials used, additive manufacturing speeds, and so on. There are many kinds of effective variables [10, 11]. The objective of this study was to investigate the stiffness or strength of unit cell models made by DMLS 3D printing. Our products need to meet defined mechanical properties for unit models to be considered for use in a sandwich core structure.

Hypercube Models

Hypercube is a mathematically ideal solution. Using a lattice or truss model in an open-cell model might have a correlation with relative elastic modulus as a function of relative density or relative compressive yield strength as a function of relative density. Figure 1 shows shapes of both Type1 and Type 2 models. Details of the stiffness and strength of both models are shown in the next section. Type 1 was a core-filled model while Type 2 was a core-spaced model. Each model was created by DMLS technique. Both models were tested under compression. Before experimental testing for Type 1 and Type 2, material properties of specimens were checked based on ASTM E8/E8M [12]. The specimen is made by DMLS and then tested under tension and compression to check the material properties (Fig. 2).

Fig. 1 Hypercube models: (1) Type 1 and (2) Type 2

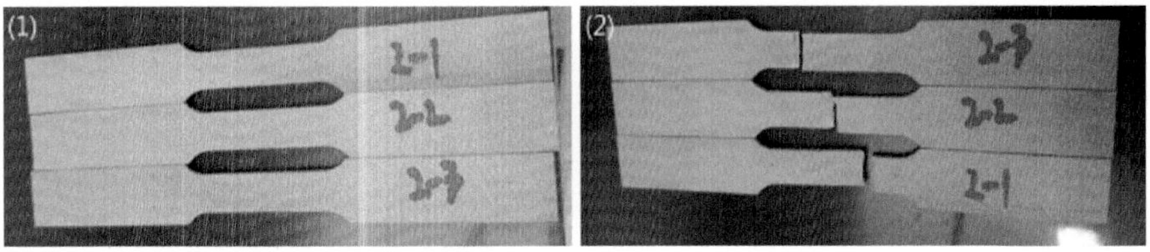

Fig. 2 Specimens made by AlSi10Mg for tension testing: (1) Specimens before tensile test; (2) Specimens after tensile test

Table 1 Mechanical properties of AlSi10Mg powder

Mechanical properties of the parts			
		As built	Heat treated
Tensile strength (MPa)	In horizontal direction (XY)	460 ± 20	345 ± 10
	In vertical direction (Z)	460 ± 20	350 ± 10
Yield strength (MPa)	In horizontal direction (XY)	270 ± 20	230 ± 20
	In vertical direction (Z)	240 ± 20	230 ± 20
Modulus of elasticity (GPa)	In horizontal direction (XY)	75 ± 10	70 ± 10
	In vertical direction (Z)	70 ± 10	60 ± 10
Elongation at break (%)	In horizontal direction (XY)	9 ± 2	12 ± 2
	In vertical direction (Z)	6 ± 2	11 ± 2

Equipment and Material Properties (Powder)

The 3D printing machine used was an EOS M290. Properties of specimens are summarized in Table 1. Detailed information about the equipment is also shown in Table 1. Among various factors, building volume and power supply for the laser are the most important factors. The maximum volume of the model was 250 mm × 250 mm × 325 mm. To make a good quality model, the volume must be at least 200 mm × 200 mm × 300 mm. The laser used was a Yb fiber laser of 40 W with a scanning speed up to 7.0 m/s, a current of 32 Amperes, and a power supply of 400 V. More information about the equipment such as dimensions, weight, software, materials used, and so on is described in Table 1. It should be noted that materials used in 3D printing are developing continuously. Nowadays, materials such as aluminum alloy, cobalt chrome, nickel alloy, stainless steel, and titanium alloy are regularly used in a powder form. Table 1 shows technical data for the aluminum used in the present study. The particular casting aluminum alloy/powder type used was AlSi10Mg. Based on ASTM E8M, specimens for compression and tension tests were made by DMLS. Figure 3 shows a specimen used in a tension test. Table 2 shows dimensions used to make tensile specimens with AlSi10Mg powder. This material has a combination of good thermal properties and low weight. These specimens were made based on the ASTM E8M standard.

Mechanical Test

Tension

Table 2 shows detailed dimensions of specimens used in tensile tests. G is a gauge length, W is width, T is thickness, R is radius of fillet, L is overall length, A is length of reduced section, B is length of grip section, and C is an approximate

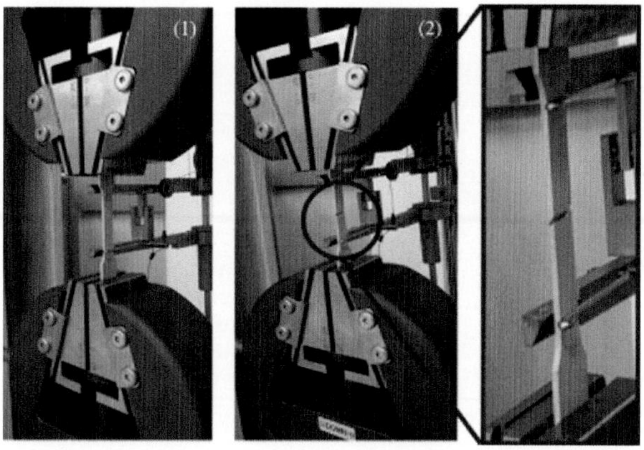

Fig. 3 Uniaxial tensile test with specimen made by AlSi10Mg at a testing speed of 1 mm/min: (1) A specimen before the tensile test; (2) A specimen after the tensile test

Table 2 Dimensions of standard specimens based on ASTM E8M

Dimensions standard specimens, sheet-type (mm)	
G - Gauge length	50.0 ± 0.10
T-Thickness	3.0
R-Radius of fillet, min	12.5
L-Overall length, min	200
A-Length of reduced section, min	57
B-Length of grip section, min	50
C-Width of grip section, approximate	20

width of grip section. For tension tests, three samples made by aluminum alloy (AlSi10Mg) were denoted as 2-1, 2-2, 2-3. Uniaxial tensile test was carried out as shown in Fig. 3. AlSi10Mg images in Fig. 2(2) are specimens after the test. Specimen 2-2 was clearly broken in the middle area. Other specimens were broken in top or bottom areas. This test revealed that 2-2 specimen was the best performing specimen. Figure 3(1) shows setting of the equipment with a mechanical extensometer and a grip. Figure 3(2) shows tested specimen 2-2 with a fracture in the middle area. From the tension test, engineering stress-strain and true stress-strain were determined as shown in Fig. 4. Comparison between engineering stress-strain and true stress-strain for the aluminum alloy AlSi10Mg is shown in Fig. 4. According to the tension test, material properties of AlSi10Mg were: a Young's modulus of 71.81 GPa, a yield strength of 155.52 MPa, and an ultimate tensile strength of 348.32 MPa. These results are summarized in Table 3. When the Young's modulus value is compared to that shown in Table 1, there were reasonable agreements for heat treated powder and in the vertical direction. Thus, specimens with required tensile strength in the vertical direction could be created by DMLS with heat treatment.

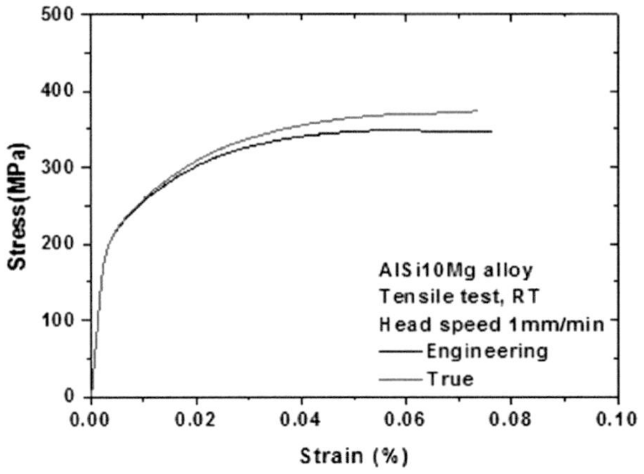

Fig. 4 Stress as a function of strain for aluminum alloy specimen after tensile test

Table 3 Material properties of AlSi10Mg powder

Material	Property	Value
Aluminum alloy, AlSi10Mg	Young's modulus (GPa)	71.8
	Yield strength (MPa)	155.5
	Ultimate tensile strength (MPa)	348.3
	Elongation (%)	8.0

Compression

Specimens used in the uniaxial compression test made by DMLS are shown in Fig. 5. We used an EOS M290 with AlSi10Mg powder. Samples were designed to have a

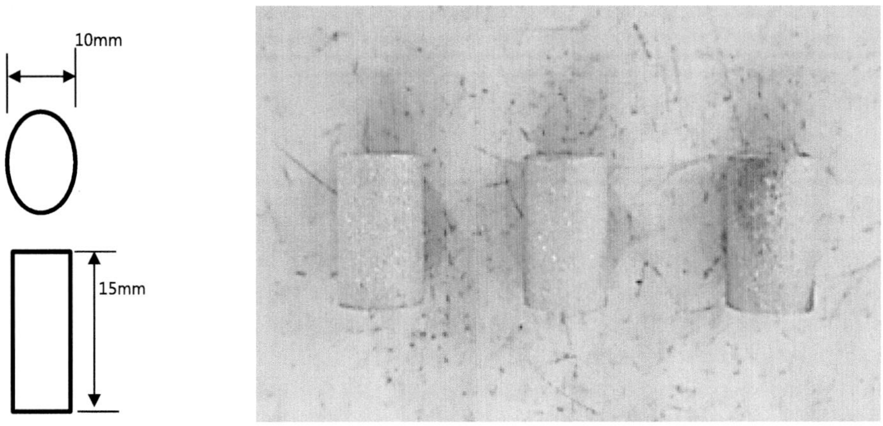

Fig. 5 Design details and AlSi10Mg specimens used for the compression test

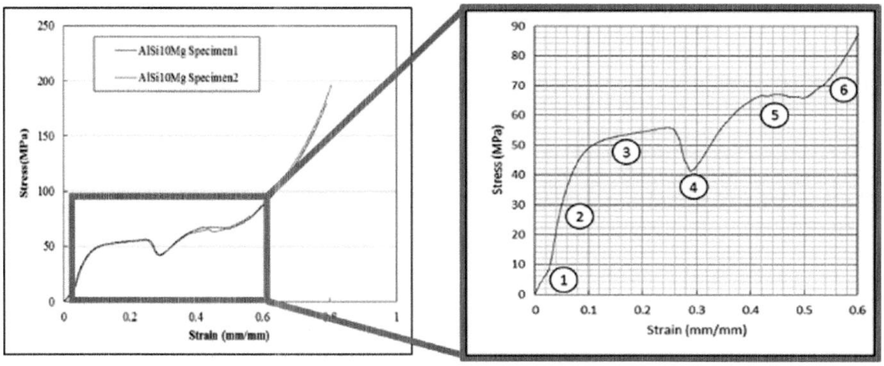

Fig. 6 AlSi10Mg Specimens for uniaxial compression test and Crushing steps: circle number 1 = elastic range, circle number 2 = linear, circle number 3 = 1st plateau, circle number 4 = valley, circle number 5 = 2nd plateau, circle number 6 = densification

diameter of 10 mm and a height of 15 mm as shown in Fig. 5. Three specimens were made for each test.

Two specimens were used for the compression test because samples matched well in the Stress-Strain plot as shown in Fig. 6. Material properties found in the compression test were: a Young's modulus of 0.316 GPa, a compressive yield strength of 6.35 MPa, and an ultimate compressive strength of 179.72 MPa.

Figure 6 shows six crushing steps with 0 to 0.6 strains. We defined these steps based on the circle number: circle number 1 = elastic range, circle number 2 = linear, circle number 3 = 1st plateau, circle number 4 = valley, circle number 5 = 2nd plateau, circle number 6 = densification. These steps were defined by Young's modulus in steps 1 and 2. We saw the 1st plateau in Step 3. From here, loading was slowly increased. The valley in step 4 appeared abruptly due to the belly phenomenon that occurred in the middle of the specimen with a height of 15 mm. This meant that the applied loading on the specimen exceeded its endurance. Then we saw the 2nd plateau in step 5 when the applied load was slowly decreased. This meant that the applied stress was distributed in the specimen. Finally, in step 6, we saw densification.

Mechanical Test for Hypercube Models

Figure 1 shows two samples made by 3D printing with DMLS. Each sample was designed to have a width of 20 mm, a length of 20 mm, a height of 20 mm, an inner truss diameter of 3 mm, and an outer truss radius of 1.5 mm. Type 1 was a core-spaced model while Type 2 was a core-filled model. Before the uniaxial compression test, measured weights for Type 1 and Type 2 models were 5.91 and 5.29 grams, respectively. The difference in weight between Type 1 and Type 2 model was about

11.72%. The material used was AlSi10Mg powder. The machine used for testing was a universal test machine (UTM) as shown in Fig. 7. Speeding was set to be 2 mm per minute. Type 1 was used in test 3 and test 4. Type 2 was used in test 1 and test 2. Figure 8 shows all tested data of stress versus strain.

For Type 1, average values of its material properties were: elastic modulus of 0.82 GPa, compressive yield strength of 2.57 MPa, ultimate compressive strength of 12.44 MPa, and compressibility of 69%. For Type 2, its material properties were: elastic modulus of 0.69 GPa, compressive yield strength of 2.29 MPa, ultimate compressive strength of 3.15 MPa, and compressibility of 46%. Material properties of

Fig. 7 A universal test machine (UTM) with data collecting system

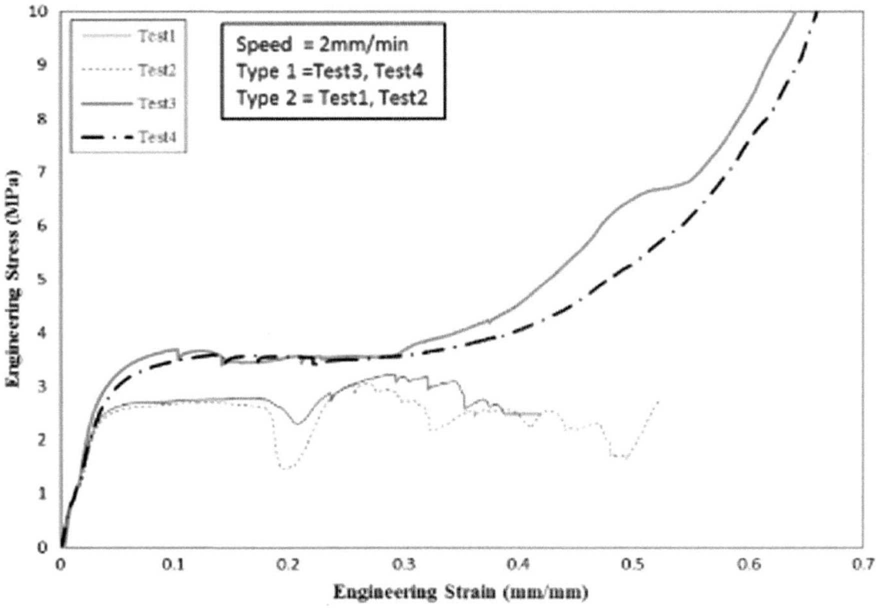

Fig. 8 Engineering stress as a function of engineering strain from uniaxial compression

Table 4 Mechanical properties of Type 1 (Core-filled model) and Type 2 (Core-spaced model)

Type	Sample number	Mechanical properties			
		Elastic modulus (GPa)	Compressive yield strength (MPa)	Ultimate comp. strength (MPa)	Compressibility (%)
1	Test 3	0.92	2.60	13.40	70
	Test 4	0.72	2.54	11.47	68
	AVG.	0.82	2.57	12.44	69
2	Test 1	0.71	2.29	3.22	42
	Test 2	0.66	2.29	3.08	49
	AVG.	0.69	2.29	3.15	46
Difference (%)		19.7	12.2	294.8	51.6

core-filled and core-spaced models are summarized in Table 4. Young's modulus, compressive yield strength, ultimate compressive strength, and compressibility of the core-filled model were 19.7%, 12.2%, 294.8%, and 51.6% higher than those of the core-spaced model (Table 4).

Conclusions

In the present study, material properties of Type 1 (a core-filled model) and Type 2 (a core-spaced model) models created by 3D printing Direct Metal Laser Sintering (DMLS) technique using aluminum alloy AlSi10Mg powder were determined. After the uniaxial compressive test, the core-filled model was found to have higher elastic modulus, compressive yield strength, ultimate strength, and percentage of elongation than the core-spaced model. These two models could be used to make a sandwich core structure. Their structures need to be investigated more deeply in the future. With improvement in 3D printing techniques, they might be useful for making aerospace materials.

Acknowledgements This research was supported by a grant (NRF-2018R1D1A1B07041383) of the Basic Science Research Program through the National Research Foundation (NRF) funded by the Ministry of Education, Republic of Korea.

References

1. Harary F (1988) A survey of the theory of hypercube graphs. Comput Math Appl 15(4):277–289
2. Mane SA (2018) Structure connectivity of hypercubes. AKCE Int J Graphs Comb 15:49–52

3. Gibson LJ, Ashby MF (1997) Cellular solids—structure and properties, 2nd edn. Cambridge University Press, UK
4. Song Y, Li Y, Song W, Yee K, Lee K-Y, Tagarielli V.L. (2017) Measurements of the mechanical response of unidirectional 3D-printed PLA. Mater Design.123(5):154–164
5. Gibson I, Shi D (1997) Material properties and fabrication parameters in selective laser sintering process. Rapid Prototyp J 3(4):129–136
6. Feng P, Meng X, Chen JF, Ye L (2015) Mechanical properties of structures 3D printed with cementitious powders. Constr Build Mater 93:486–497
7. Engkvist G (2017) Investigation of microstructure and mechanical properties of 3D printed Nylon, Master thesis. Lulea University of Technology, Materials Engineering Master's level
8. Basgul C, Yu T, MacDonald DW, Siskey R, Marcolongo M, Kurtz SM (2018) Structure-property relationships for 3D–printed PEEK intervertebral lumbar cages produced using fused filament fabrication. J Mater Res 33(4):2040–2051
9. Kundu S, Hussain M, Kumar V, Mumar S, Das AK (2018) Direct metal laser sintering of TiN reinforced Ti6Al4V alloy based metal matrix composite: fabrication and characterization. Int J Adv Manuf Technol 97:2635–2646
10. Cantrell J, Rohde S, Damiani D, Gurnani R, diSandro L, Anton J, Young A, Jerez A, Steinbach D, Kroese C, Ifju P (2017) Experimental characterization of the mechanical properties of 3D printed ABS and polycarbonate parts. In: Yoshida S, Lamberti L, Sciammarella C (eds) Advancement of optical methods in experimental mechanics, 3, the conference proceedings of the society for experimental mechanics series book series (CPSEMS). Springer, Cham, pp 89–105
11. Leite, M, Fernandes J, Deus AM, Reis L (2018) Study of the influence of 3D printing parameters on the mechanical properties of PLA. In: 3rd international conference on progress in additive manufacturing, Pro-AM 2018, Singapore
12. ASTM E8/ E8M-13 (2013) Standard test methods for tension testing of metallic materials, ASTM International, West Conshohocken, PA

Study on Deformation of Closed-Cell Aluminum Foam in Different Solid–Liquid–Gas Coexisting States

Zhiyong Liu, Ying Cheng, Yanxiang Li, Ningzhen Wang and Xu Zhou

Abstract The deformation of closed-cell aluminum foam (CAF) in different solid–liquid–gas coexisting states was studied in this paper. The results showed that the most suitable temperature of the CAF deformation was about 635 °C, which made the CAF have maximum deformation, but its characteristic parameters changed slightly. Moreover, the multi-grain cell wall and small size cell were helpful to the CAF deformation in the solid–liquid–gas coexisting state. When the CAF was heated at 635 °C, the intergranular solidification microstructure and part of the primary α-Al grain were melted into liquid. The fixed or inter-locked primary α-Al grain was released and changed into the distributed discretely near-spherical grain, making the cell wall have good thixotropy. At the same time, the gas pressure in the cell was recovered to the foaming stage, so the gas pressure could be close to the flow stress of cell wall. During the CAF deformation in the solid–liquid–gas coexisting state, the cell wall deformed in thixotropic under the supporting and coordinating of the gas pressure in the cell; meanwhile, the cell was moved following the cell wall deformation. Therefore, the CAF deformation was affected by the types of cell walls and cell size.

Keywords Closed-cell aluminum foam · Deformation temperature · Special shaped · Solid–liquid–gas coexisting state

Z. Liu (✉)
School of Mechanical and Electrical Engineering, Heze University, Heze 274015, Shandong, China
e-mail: yyaffe@126.com

Z. Liu · Y. Cheng · Y. Li · N. Wang · X. Zhou
School of Materials Science and Engineering, Tsinghua University, Beijing 100084, China

Z. Liu
School of Engineering, Wayne State University, Detroit, MI 48202, USA

N. Dukhan (ed.), *Proceedings of the 11th International Conference on Porous Metals and Metallic Foams (MetFoam 2019)*, The Minerals, Metals & Materials Series, https://doi.org/10.1007/978-3-030-42798-6_7

Introduction

Closed-cell aluminum foam(CAF) has unique physical and mechanical properties, which made it have broad application prospects [1, 2]. The preparation technology and mechanical properties of the CAF have attracted wide attention. Although the preparation technology has been mastered [3, 4], the mechanical properties are still to be studied further. At present, almost all of the researches in the mechanical properties of the CAF mainly focus on failure and destruction [5–8], rather than effective deformation. As a result, the special shaped CAFs develop slowly [9]. Because the CAF has poor mold filling ability and cell stability in the liquid state and the collapse will occur layer by layer when the CAF deforms in solid state [10–13], the special shaped CAFs can't be produced through casting and stamping. What's more, the special shaped CAFs can't be obtained by welding, as its isotropy will be destroyed [14–16]. Although the machining, the powder metallurgy foaming, and the two-step foaming method can be used to manufacture the special shaped CAF [17–19] , there are some shortcomings, such as the lower utilization rate of materials, or strict and complex process. In order to develop the special shaped CAF, the CAF deformation in solid–liquid–gas coexisting state has been proposed [20].

The preparation of CAF is similar to the preparation of semi-solid metal to certain degree [21]. When the CAF is heated to the semi-solid temperature of cell wall, the CAF is in solid–liquid–gas coexisting state, which makes the cell wall have good thixotropy [22] and makes the gas pressure in cell increase because the gas was closed in cells. The CAF deforms in solid–liquid–gas coexisting state through the thixotropic deformation of the cell wall and the accompanying motion of cell. The authors consider that under a certain temperature the gas pressure in the cell can be close to the flow stress of the cell wall [23], which is helpful to keep the continuity of the cell wall and the integrity of the cell during the CAF deformation. The flow stress of the cell wall is related to liquid fraction and character distribution of primary α-Al grain, which is affected by deformation temperature. Therefore, the effect of the temperature on the CAF deformation in solid–liquid–gas coexisting state is studied in this paper.

Experiments

Materials

The CAF used here was prepared by melt foaming process [24]. The relative density of CAF was 33.6%, the average equivalent diameter of the cell was 1.39 mm, and the alloy composition of the cell wall (wt%) was Ca 3.0, Ti 0.02, and Al balanced. The solid–liquid temperature range of the cell wall was measured by NETZSCH STA449 F3 differential scanning calorimeter. The result showed that approximate solid–liquid

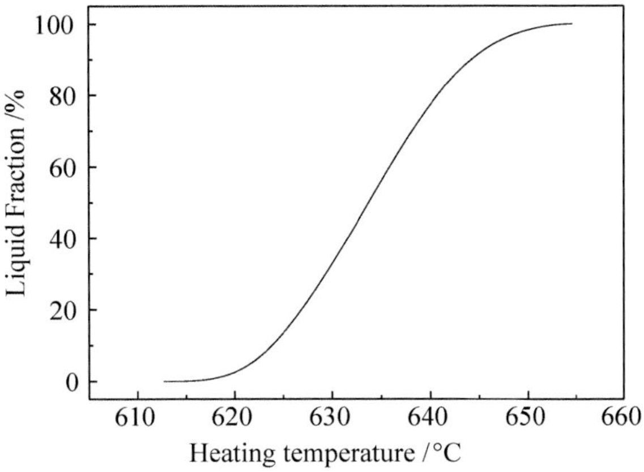

Fig. 1 Curve of liquid fraction and temperature of CAF

temperature range of the cell wall was about 612–657 °C, and relationship curve between liquid fraction and temperature is shown in Fig. 1.

Methods

Firstly, the CAF plate without size effect ($\Phi 65 \times h8$ mm) [25, 26] was placed on the joint surface of the female mold, and a cross groove was designed at bottom of the female mold, as shown in Fig. 2a. Both the CAF and molds were heated in resistance furnace. The heating rate of resistance furnace was 10 °C/min, and the heating process was in an inert atmosphere. The heating temperature and deformation temperature were measured and controlled by Ni–Cr/Ni–Si thermocouples. When the CAF was heated to schedule temperature, the CAF plate was deformed immediately through mold pressing. The pressing speed of male mold was about 0.5 mm/s, and

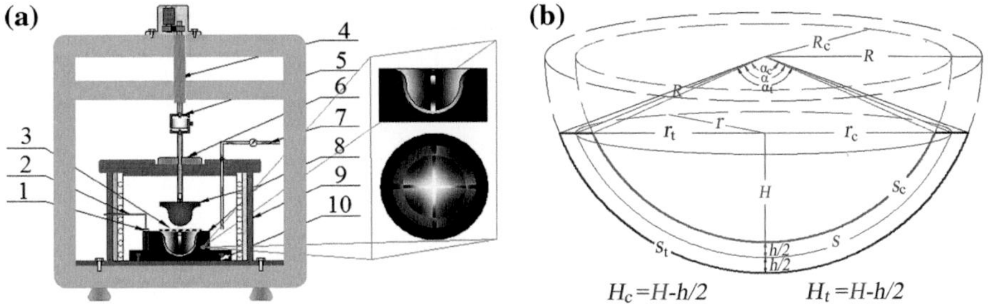

Fig. 2 Deformation schematic diagram and geometric schematic diagram of deformed CAF. 1-Temperature-controlled CAF; 2-Thermocouple; 3-CAF plate; 4-Power device; 5-Pressure sensor; 6-High-temperature seal ring; 7-Inert gas; 8-Graphite mold; 9-Crucible furnace; 10-Limit guide rails

Table 1 Process parameter, macro-deformation, and characteristic parameter of the CAF

PARM. Samp.	Process parameters		Macro-deformation		Characteristic parameters		
	T※/ °C	σ^*/MPa	ε_c/%	ε_t/%	D/mm	F	ρ_r #/%
A	–	–	–	–	1.39	0.93	33.6
B	615	1.02	2.7	6.7	1.38	0.90	32.4
C	625	0.90	3.2	9.1	1.36	0.88	31.8
D	635	0.51	4.0	16.5	1.34	0.85	33.1
E	645	0.43	4.6	14.2	1.29	0.81	34.0
F	655	0.31	10.1	11.6	1.23	0.82	34.8

T※Deformation temperature; σ^* Deform pressure; ρ_r #Relative density

the maximum pressure was recorded by pressure sensor, as shown in Table 1. After the joint surfaces came into contact with one another, the mold pressure was kept for 5 s before releasing load, and the spherical crown was obtained. Subsequently, the spherical crown and female mold were moved out immediately from the resistance furnace through guide rails, and they were cooled to room temperature by 0.1 L/min compressive argon.

And then, the macroscopic morphology and deformation of the spherical crown were studied by macroscopic observation and calculation, respectively, so the formula (1) was introduced. The mesoscopic deformation of the CAF was analyzed through using the SG3100 three-dimensional optical microscope and Image-pro plus software. The average equivalent diameter D and shape factor F of cells were calculated by formulas (2) and (3), respectively [20]. The microstructure evolution of CAF was analyzed by Neuphoto 21 metallurgical microscope and Image-pro plus software.

$$\varepsilon_{i(i=t,c)} = \frac{\left|\arctan \frac{2(R_i - H_i)r_i}{R_i^2 - 2r_i^2} R_i - \phi\right|}{\phi} \times 100\% \quad (1)$$

where ε_i, R_i, H_i, r_i, and Φ are deformation of the CAF, radius of sphere, height of spherical crown, radius of spherical crown, and bottom radius of the spherical crown, respectively, as shown in Fig. 2b.

$$D = \sum_{i=1}^{N} \frac{\sqrt{4A/\pi}}{N} \quad (2)$$

$$F = \sum_{i=1}^{N} \frac{4\pi A/p^2}{N} \quad (3)$$

where *D*, *F*, *A*, *N*, and *p* are the average cell size, shape factor, area, number, and perimeter of the two-dimensional cell, respectively.

Finally, the CAFs which were the same as the deforming CAF were heated to the corresponding deformation temperature, water-cooled immediately, to study the characteristic distribution of Al_4Ca compound. The ZEISS-SUPRA40 scanning electron microscope was used, and the surface scan of Ca element was observed.

Experimental Results

As shown in Table 1, when the CAF deformed in different solid–liquid–gas states, its macroscopic deformation was quite different. With the deformation temperature increasing, the compress deformation gradually enhanced, and increased sharply above 645 °C. However, the tensile deformation firstly increased and then decreased with the deformation temperature increasing, so the maximum tensile deformation was obtained at 635 °C, as shown in Fig. 3a. In addition, the mesoscopic deformation also went through the similar variation. With increasing deformation temperature,

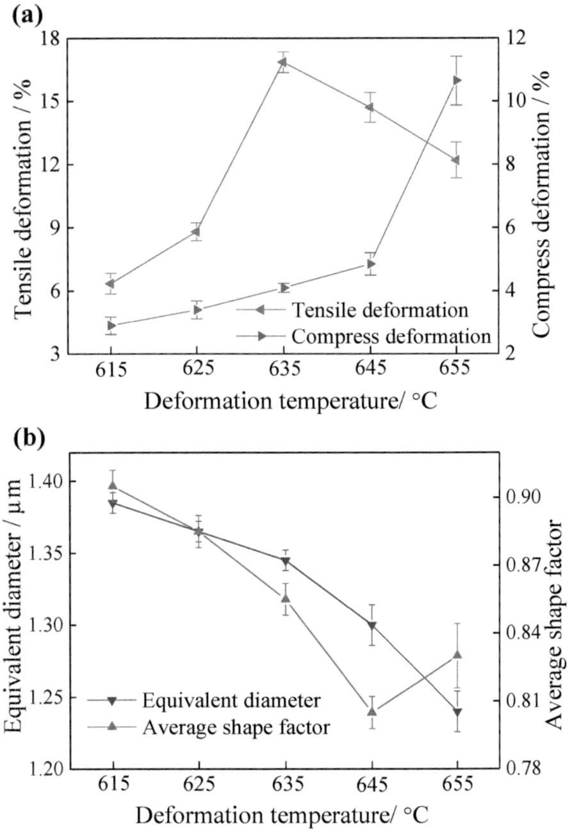

Fig. 3 Curves of deformation versus deformation temperature. **a** Macroscopic deformation; **b** Mesoscopic deformation

the average equivalent diameter continually decreased, and it decreased slowly under 635 °C, but sharply above 635 °C. However, the shape factor firstly decreased and then increased with the deformation temperature increasing, so the minimum shape factor was obtained at 645 °C, as shown in Fig. 3b.

Macroscopic Deformation of the CAF

The macroscopic morphology of the initial and deformed CAF is shown in Fig. 4. When the CAF deformed under 635 °C, it had lower liquid fraction and needed higher deformation stress, as shown in Fig. 1 and Table 1. The deformation form of the CAF

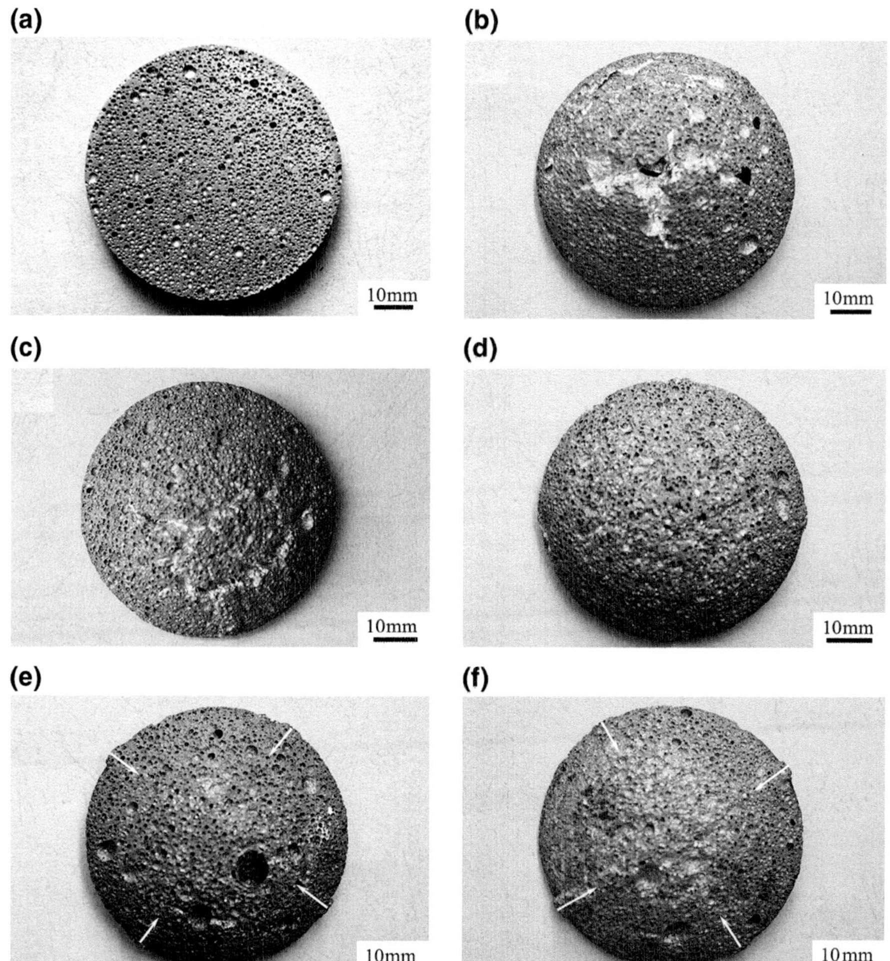

Fig. 4 Macroscopic morphology of initial and deformed CAFs. **a** Initial CAF; **b** 615 °C; **c** 625 °C; **d** 635 °C; **e** 645 °C; **f** 655 °C

was high-temperature deformation and was sensitive to large size cell and cell wall defect [27, 28], which made local fractures occur near the large size cell. Therefore, the compress and tensile deformation of the CAF were lower. Remarkably, with the deformation temperature increasing, the local fracture form changed from plastic fracture to solid–liquid separation, as shown in Fig. 4b, c. When the deformation temperature increased to 635 °C, the local fracture caused by large size cell and cell wall defect disappeared completely, making the CAF deform uniformly, and the maximum tensile deformation was obtained, as shown in Fig. 4d. However, when the deformation temperature further increased above 635 °C, the CAF showed fluid characteristic, and the local collapse caused by large size cell and cell wall defect was gradually obvious. Therefore, there were several raised points or ridges at the top of the deformed CAF corresponding to the cross groove, and the compress deformation sharply increased when the deformation temperature was above 645 °C, as shown in Figs. 3a and 4e, f.

Mesoscopic Deformation of the CAF

The mesoscopic morphology of the initial and deformed CAF is shown in Fig. 5. The net-like structure was the cell wall, and the mesh area surrounded by cell wall was the cell in 2D space. Before the CAF deformed, the cell wall was smooth and the cell was near-sphere, as shown in Fig. 5a. When the CAF deformed under 635 °C, most of the cell wall structure, the cell morphology, and the spherical microcell were maintained. The average diameter and factor shape of cell changed slightly, as shown in Fig. 3b. However, there were several fractured cell walls making the integrality of cell unit destroyed. It is noteworthy that with the deformation temperature increasing, the fracture of cell wall weakened obviously, as shown in Fig. 5b, c. When the deformation temperature increased to 635 °C, the fracture of cell wall disappeared basically and some cell walls changed from smooth to concave–convex, resulting in the average diameter and factor shape of cell decreasing, the cell morphology changed from near-sphere or sphere to polyhedron after the cell wall deforming under the support of gas pressure, as shown in Fig. 5d. When the CAF deformed above 645 °C, the cell wall structures and cell morphology changed seriously, as shown in Fig. 5e, f. In this case, the CAF had high liquid fraction and showed obvious fluid characteristics. Therefore, the cell easily ruptured, and as a result, the cell walls were easily squeezed together, and with the temperature increasing, the cell walls even merged into new cell walls and cell units, which made compress deformation and relative density increasing sharply; meanwhile, the shape factor increased slightly. The analysis mentioned above could explain why the factor shape improved slightly at 655 °C.

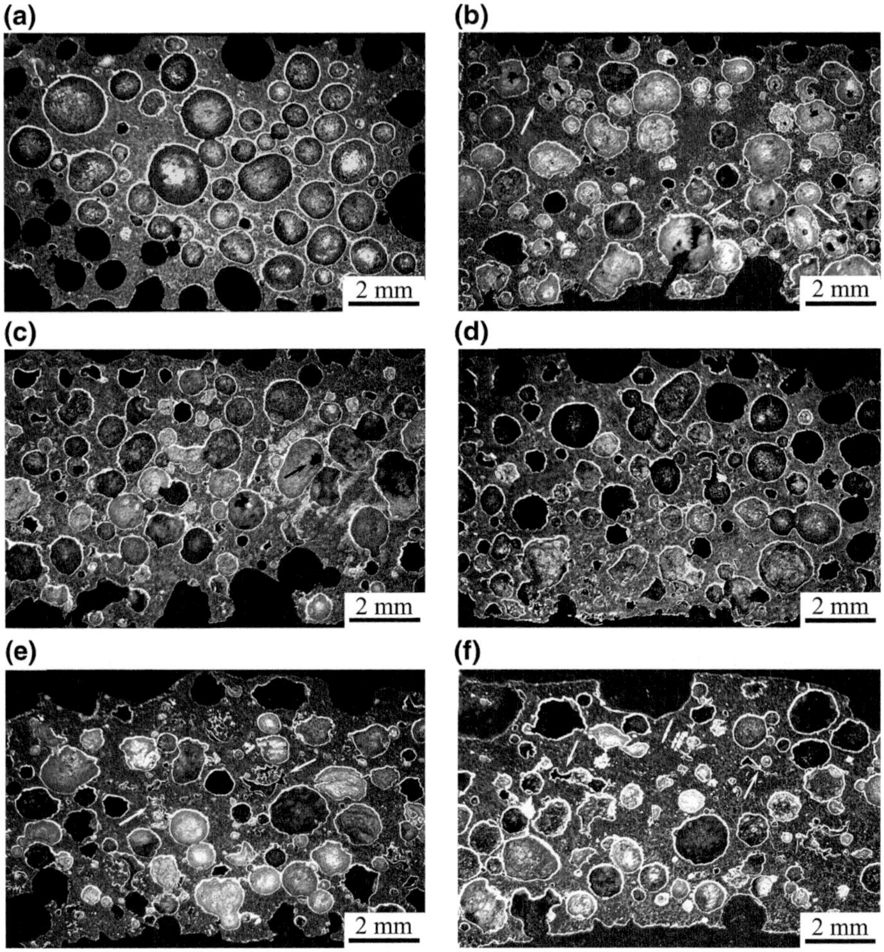

Fig. 5 Mesoscopic morphology of initial and deformed CAFs **a** Initial CAF; **b** 615 °C; **c** 625 °C; **d** 635 °C; **e** 645 °C; **f** 655 °C

Microstructure Evolution of the CAF

The microstructure of the initial and deformed CAFs is shown in Fig. 6. The net-like structure was the cell wall, and the mesh area surrounded by cell wall was the cell in 2D space. The cell wall consisted of white particulate primary α-Al grain, the gray intergranular solidification structure, and spherical microcell. In the initial CAF, the particulate primary α-Al grain and intergranular solidification structure were mainly degenerated dendrite and Al–Al_4Ca pseudo eutectic, as shown in Fig. 6a. When the CAF deformed under 635 °C, the primary α-Al grain and intergranular solidification structure changed rarely, and the fine dendrite protrusion and sharp

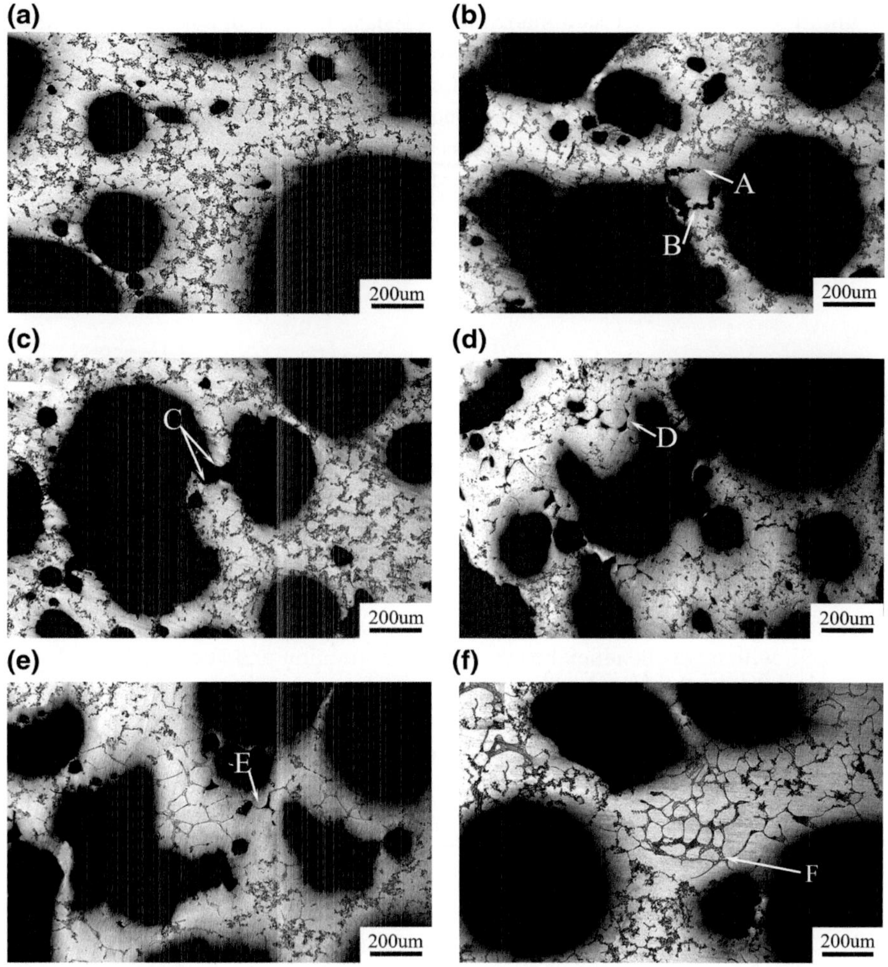

Fig. 6 Microstructures of initial and deformed CAFs. **a** Initial CAF; **b** 615 °C; **c** 625 °C; **d** 635 °C; **e** 645 °C; **f** 655 °C

corner were largely maintained. However, the fracture form of cell wall changed from mixed of transgranular and intergranular to intergranular with the deformation temperature increasing, as shown by arrows "A", "B", and "C" in Fig. 6b, c, which was consistent with solid–liquid separation that showed in macroscopic morphology. The change of microfracture mechanism made the deformation increase and the relative density decrease. When the deformation temperature increased to 635 °C, the intergranular solidification structure was melted completely and primary α-Al grain got ripened and melted partially. Meanwhile, some ultrafine irregular microcells formed in intergranular region, which meant that the primary α-Al grain moved or deformed with the liquid flowing, as shown by arrow "D" in Fig. 6d. Moreover, the ultrafine irregular microcell made the CAF have maximum tensile deformation. When the deformation temperature was above 645 °C, the primary α-Al grain got ripened sufficiently, and the number of the ultrafine irregular microcell decreased, which was helpful to the cell wall deformation, as shown by arrow "E" in Fig. 6e. But when the deformation temperature was too high, most of the primary α-Al grains were melted, the CAF deformation was similar to thixoforming of the semi-solid alloy with the supporting of gas pressure in cell. After the CAF was deformed at 655 °C, the intermetallic compounds redistribute among the primary grains and secondary grains, as shown by arrow "F" in Fig. 6f.

Discussion

When the CAF was in a specific solid–liquid–gas coexisting state, the gas pressure in the cell was recovered largely to the foaming stage and the intergranular solidification structure and sharp corner of primary α-Al grain were melted into liquid, which made the fixed or inter-locked primary α-Al grain released. At the same time, the primary α-Al grain size decreased and the distance between primary α-Al grains increased. Therefore, the difference between gas pressure in cell and flow stress of cell wall decreased, which was beneficial to the CAF deformation and keeping the cell wall structure and cell morphology. However, the analysis above was only applicable to ideal CAF, in fact, the CAF was composed of the complex cell units [29].

The cell wall of CAF was categorized into single-grain cell wall and multi-grain cell wall depending on primary α-Al grain size in cell wall and thickness of the cell wall [30]. Therefore, the CAF deformation was composed of several deformation forms. When cell unit had single-grain cell wall(s), deformation of the cell unit might include high-temperature deformation or solid–liquid separation. When the single-grain cell wall was fixed by adjacent primary α-Al grain, the single cell wall would deform in high-temperature form. But if there were enough space and degree of freedom between single-grain cell wall and its adjacent liquid, the single cell wall might move out from its adjacent liquid, namely, solid–liquid separation. Either high-temperature deformation or solid–liquid separation would make the cell wall structure and cell morphology destroyed, so single-grain cell wall was bad for the CAF deformation. When cell unit was composed of multi-grain cell wall(s), its deformation included the thixotropic deformation of cell wall and the accompanying motion of cell. The deformation stress of the CAF made primary α-Al grain move or deform the following liquid flowing in cell wall. As a result, the relative motions between the primary α-Al grains and between the primary α-Al grain and liquid occurred, that is, cell wall deformed in thixotropic, which made the cell move accordingly. The deformation of multi-grain cell wall and the gas pressure in cell affected each other, and they also were affected by the deformation temperature and radius of cell. Therefore, The deformation of multi-grain cell wall was a complex process and had a dominant effect on the CAF deformation.

The deformation of multi-grain cell wall was closely related to its microstructure. The microstructure of multi-grain cell wall such as the liquid fraction and characteristic distribution of primary α-Al grain was affected by deformation temperature. The space and degree of freedom required for primary α-Al grain moving and deforming mainly depended on the liquid fraction and characteristic distribution of primary α-Al grain. When the CAF deformed under 635 °C, it had lower liquid fraction and its primary α-Al grain was fixed and centrally distributed, making primary α-Al grain move difficultly. So both the compress deformation and tensile deformation only changed slightly. When the deformation temperature increased to 635 °C, the primary α-Al grain got ripened or even partially melted, which made itself round and small and made the distance between primary α-Al grains increase. The probability of collision and friction between primary α-Al grains decreased, which made the possibility of the relative motion between primary α-Al grains improve. However, during the relative motion between primary α-Al grains, liquid easily adhered to primary α-Al grain instead of filling bridging gap. The bridging gap was formed by point contact caused by relative motion between primary α-Al grains. As a result, ultrafine irregular microcell was formed in intergranular region, which not only made the cell wall structure and cell morphology change significantly, but also made the CAF have maximum tensile deformation. When the deformation temperature increased to above 645 °C, most of the primary α-Al grains were melted, resulting in the liquid fraction increasing sharply. The irregular microcell formed because of the bridging gap decreased or even disappeared. Moreover, the flow stress of cell wall decreased sharply, which made the cell units easily collapse, so the compress deformation increased sharply.

The deformation of multi-grain cell wall was also closely related to gas pressure in cell. When the CAF deformed in solid–liquid–gas coexisting state, the gas pressure in cell played a supporting and coordinating role in the cell wall deformation. The gas pressure in cell mainly depended on deformation temperature and compress degree, and the gas pressure could be close to the flow stress of multi-grain cell wall in specific condition [31]. According to the Bernoulli equation and rheological theory, the gas pressure in cell and flow stress of multi-grain cell wall were expressed by formulas (4), (5), and (6), respectively [32].

$$\frac{PV}{T} = \frac{nP(V - V')}{T_d} \tag{4}$$

$$V' = \frac{nPV - PV}{nP} = \frac{4(n-1)\pi R^2}{3n} \tag{5}$$

$$\tau = 2.79 \times \exp(6.27 \times \frac{T_L - T_d}{T_L - T_S}) \times \gamma^{(1-0.7)} \tag{6}$$

where P, T, T_d, n, V', V, R, τ, T_L, T_S, and γ are gas pressure in cell at room temperature, room temperature, deformation temperature, multiples coefficient, compressed volume of cell, volume of initial cell, radius of cell, flow stress of cell wall, liquidus temperature, solidus temperature, and shear rate, respectively.

The formulas above showed that the lower the deformation temperature was, the lower the gas pressure in cell was, but the flow stress of cell wall was larger. In theory, when the deformation temperature was low, only by compressing the cell seriously could its gas pressure be close to flow stress of the cell wall. In fact, when the deformation temperature was under 635 °C, the flow stress of the cell wall was very high. Even under high deformation stress, it was difficult for the cell unit to be compressed, or else the cell would be destroyed. Therefore, the deformation and characteristic parameters of CAF nearly unchanged. When the deformation temperature increased to 635 °C, the flow stress of cell wall decreased obviously, so the deformation and characteristic parameters of the CAF started to change obviously. When the deformation temperature increased to above 635 °C, the flow stress decreased sharply, but the gas pressure increased sharply. As a result, the characteristic parameters of CAF changed significantly and some cell units even collapsed, and the compress deformation increased sharply.

The formula (5) also showed that the compress degree of cell was closely related to the radius of cell. If the compressed part of cell was a spherical cap in shape, the relationship between radius of cell and compressed height was shown in formula (7) whose derivative was formula (8). Therefore, the surface area of spherical cap increased with the radius increasing, as shown in formula (9).

$$\frac{4(n-1)\pi R^2}{3n} = \frac{\pi h^2}{2}(2R - h) \tag{7}$$

$$\frac{dh}{dR} = \frac{h^2}{R} \cdot \frac{2R - h}{R^2 - (h - R)^2} \tag{8}$$

$$S = 2\pi Rh \tag{9}$$

where h, R, and S are compress height, radius of cell, and surface area of spherical cap, respectively. In addition, h varied in the range of $0 \leq h < 2R$.

Formula [5, 9] showed that, when the different size cell units got the same gas pressure under the same specific deformation temperature, the compress degree of large size cell units was more serious. During the CAF deforming in solid–liquid–gas coexisting state, the thixotropic deformation of the cell wall was supported by the gas pressure in cell and coordinated by the gas pressure in adjacent cells. The cell moved accompanied by the cell wall deformation, but only when the gas pressure in adjacent cells was similar would the adjacent cells move simultaneously. Therefore, the deformation capacity of CAF might be improved by decreasing the cell size and improving the uniform of the cell.

Conclusions

(1) Under the condition that the macroscopic morphology, mesoscopic morphology, and microstructure are considered comprehensively, the CAF has good deformation capacity at about 635 °C.
(2) The multi-grain cell wall is beneficial to the deformation in the solid–liquid–gas coexisting state. Besides, the deformation capacity of the CAF can be improved by increasing the number of multi-grain cell walls and improving the characteristic distribution of the primary α-Al grain.
(3) During the CAF deforming in solid–liquid–gas coexisting state, multi-grain cell wall deforms in thixotropic, which is supported by the gas pressure in cell and coordinated by the gas pressure in adjacent cells. The deformation capacity of the CAF can be increased by decreasing the cell size and improving the uniform of the cell.

References

1. Banhart J (2013) Light-metal foams-history of innovation and technological challenges. Adv Eng Mater 15(3):82–111. https://doi.org/10.1002/adem.201200217
2. Patel P, Bhingole PP, Makwana D (2018) Manufacturing, characterization and applications of lightweight metallic foams for structural applications: review. Mater Today Proc 5(9):20391–20402
3. Zhou Y, Li Y, Yuan J (2015) The stability of aluminum foams at accumulation and condensation stages in gas injection foaming process. Colloids Surf A 482(5):468–476
4. Banhart J (2018) Production of metal foams. In: Beaumont PWR, Zweben CH (eds) Comprehensive composite materials. Elsevier, Oxford, pp 347–363
5. Lehmhus D, Banhart J (2003) Properties of heat-treated aluminum foams. Mater Sci Eng A 349(1–2):98–110
6. Huang R, Ma S, Zhang M, Xu J, Wang Z (2019) Dynamic deformation and failure process of quasi-closed-cell aluminum foam manufactured by direct foaming technique. Mater Sci Eng A 756(22):302–311
7. Ulbin M, Glodež S, Vesenjak M, Duarte I, Podgornik B, Ren Z, Kramberger J (2019) Low cycle fatigue behaviour of closed-cell aluminium foam. Mech Mater 133:165–173
8. Wang P, Xu S, Li Z, Yang J, Zheng H, Hu S (2014) Temperature effects on the mechanical behavior of aluminum foam under dynamic loading. Mater Sci A 599(2):174–179
9. Utsunomiya H, Matsumoto R (2014) Deformation processes of porous metals and metallic foams (Review). Procedia Mater Sci 4:245–249
10. Kader MA, Islam MA, Saadatfar M, Hazell PJ, Brown AD, Ahmed S, Escobedo JP (2017) Macro and micro collapse mechanisms of closed-cell aluminium foams during quasi-static compression. Mater Des 118(15):11–21
11. Nosko M, Simančík F, Iždinský K, Švec P, Florek R (2011) Stabilizing intermetallic phases within aluminum foam. Mater Lett 65(9):378–1380
12. Leitlmeier D, Degischer HP, Flankl HJ (2002) Development of a foaming process for particulate reinforced aluminum melts. Adv Eng Mater 4(10):735–740
13. Song Z, Ma L, Wu Z, He D (2000) Effects of viscosity on cellular structure of foamed aluminum in foaming process. J Mater Sci 35(1):15–20

14. Wan L, Huang Y, Huang T, Lv Z, Feng J (2016) Interfacial behavior and mechanical properties of aluminum foam joint fabricated by surface self-abrasion fluxless soldering. J Alloy Compd 671(25):346–353
15. Shih J, Zeng YT, Yang J (2011) Principal component analysis for multiple quality characteristics optimization of metal inert gas welding aluminum foam plate. Mater Des 32(3):1253–1261
16. Nowacki J, Moraniec K (2015) Welding of metallic AlSi foams and AlSi–SiC composite foams. Arch Civ Mech Eng 15(4):940–950
17. Chu X, Wang H, He S, He D (2009) Study on fabrication of shaped al alloy foam by two-step foaming method. Int J Mod Phys B 23(6–7):972–977
18. Geramipour T, Oveisi H (2017) Effects of foaming parameters on microstructure and compressive properties of aluminum foams produced by powder metallurgy method. Trans Nonferr Metal Soc 27(7):1569–1579
19. Hangai Y, Amagai K, Omachi K, Tsurumi N, Utsunomiya T, Yoshikawa N (2018) Forming of aluminum foam using steel mesh as die during foaming of precursor by optical heating. Opt Laser Technol 108:496–501
20. Liu Z, Cheng Y, Li Y, Zhou X, Chen X, Wang N (2018) Shape formation of closed-cell aluminum foam in solid-liquid-gas coexisting state. Int J Min Met Mater 25(8):974–980
21. Lashkari O, Ghomashchi R (2007) The implication of rheology in semi-solid metal processes: an overview. J Mater Proc Tech 182(1–3):229–240
22. Song Z, Nutt SR (2007) Rheology of foaming aluminum melts. Mater Sci Eng A 458:108–115
23. Aly MS (2007) Behavior of closed cell aluminium foams upon compressive testing at elevated temperatures: experimental results. Mater Lett 61:3138–3141
24. Cheng Y, Li Y, Chen X, Shi T, Liu Z, Wang N (2017) Fabrication of aluminum foams with small pore size by melt foaming method. Metall Mater Trans B 48(1–2):754–762
25. Zhang Y, Jin T, Li S, Ruan D, Wang Z, Lu G (2019) Sample size effect on the mechanical behavior of aluminum foam. Int J Mech Sci 15:622–638
26. Hu Y, Fang Q, Sha B, Zhao M (2018) Effect of the large cells on the fatigue properties of closed-cell aluminum alloy foam. Compos Struct 200(15):59–68
27. Sahu S, Goel MD, Mondal DP, Das S (2014) High temperature compressive deformation behavior of ZA27–SiC foam. Mater Sci Eng A 607:162–172
28. Chen Y, Das R, Battley M (2017) Effects of cell size and cell wall thickness variations on the strength of closed-cell foams. Int J Eng Sci 120:220–240
29. Jang W, Hsieh W, Miao C, Yen Y (2015) Microstructure and mechanical properties of ALPORAS closed-cell aluminum foam. Mater Charact 107:228–238
30. Islam MA, Kader MA, Hazell PJ, Brown AD, Saadatfar M, Quadir MZ, Escobedo JP (2016) Investigation of microstructural and mechanical properties of cell walls of closed-cell aluminum alloy foams. Mater Sci Eng A 666(1):245–256
31. Sadot O, Ram O, Anteby I, Gruntman S, Ben-Dor G (2016) The trapped gas effect on the dynamic compressive strength of light aluminum foams. Mater Sci Eng A 659:278–286
32. Kang CG, Yoon JH, Seo YH (1997) The upsetting behavior of semi-solid aluminum material fabricated by a mechanical stirring proces. J Mater Process Technol 66(1–3):6630–6638

Study on the Mechanical Properties of Composite Metal Foam Core Sandwich Panels

Jacob Marx and Afsaneh Rabiei

Abstract Metal foams are known for their energy absorption under compression but generally lack strength in tension. Metal foam core sandwich panels can offer a non-porous surface and improve the material's corrosion resistance as well as their strength in tension. Composite metal foam (CMF) is a type of metal foam that has superior mechanical properties compared to other metallic foams due to the presence of a matrix in between its porosities and their uniform cell structure. The mechanical properties of CMF under various loading scenarios are well established, but mechanical properties of CMF core sandwich panels (CMF-CSP) have yet to be reported. This work will report the mechanical properties of CMF-CSPs and compare those to bare CMF panel properties under compression, bending, and tension. CMF-CSPs were manufactured by diffusion bonding stainless steel face sheets to the top and bottom of stainless steel CMF plates. The tensile and bending yield strengths and failure strain saw a drastic increase due to the addition of the face sheets, while the compressive properties of the CMF core were unchanged. Overall, CMF-CSP can be preferred for applications that require a nonporous surface, such as those exposed to corrosive environment like material used in tank cars carrying hazardous materials, in machine tools, and even basic components of building structures. Alternative adhesions methods for the face sheets are currently explored and will be reported in future studies.

Keywords Composite metal foam · Sandwich panel · Mechanical testing · Compression · Bending · Tension

J. Marx · A. Rabiei (✉)
Department of Mechanical and Aerospace Engineering, North Carolina State University, 3250 Engineering Bldg. 3, 911 Oval Dr., Campus Box 7910, Raleigh, NC 27695-7910, USA
e-mail: arabiei@ncsu.edu

N. Dukhan (ed.), *Proceedings of the 11th International Conference on Porous Metals and Metallic Foams (MetFoam 2019)*, The Minerals, Metals & Materials Series, https://doi.org/10.1007/978-3-030-42798-6_8

Introduction

Metal foams are known for their unique energy absorption capabilities as well as their inherent thermal and acoustic insulating properties [1]. The use of metal foams in various applications can be limited by their strength under tension and vulnerability to corrosion due to their large surface area. The porosities in metal foams hinder their tensile strength and allow for rapid crack growth through the structure [2]. Metal foam core sandwich panels can be manufactured to create a non-porous surface and support the core structure under tension while providing a net reduction in the exposed surface, improving the corrosion resistance of the core. Composite metal foam (CMF) is proven to have superior mechanical properties compared to other metal foams due to the presence of a metallic matrix between its porosities [3–5]. CMF is manufactured by surrounding hollow metal spheres with a metallic matrix, creating a product with comparable ductility of its components and improved adherence between the porosities. The compressive behavior of CMF is well documented and understood [3–5]. In this study, the manufacturing of CMF-CSP is reported along with the mechanical and microstructural properties of the CMF-CSP. The mechanical properties of the CMF-CSP are compared to bare CMF samples under the same loading conditions.

Materials and Processing

Stainless steel composite metal foam (CMF) panels were manufactured using a powder metallurgy technique described in prior works [3]. Stainless steel spheres used for processing the CMF panels were manufactured by Hollomet GmbH located in Dresden, Germany and had an average outer diameter of 2 mm with a wall thickness of 100 μm. The spheres were surrounded with 316L stainless steel powder from North American Höganas with an average particle size of 44 μm. The panels were manufactured in 25 $\times$ 25 cm with a variety of thicknesses designed to meet the required ASTM standards for cutting compression, tensile, and bending samples. The resulting CMF samples had a density of 2.95 g/cm^3. Two CMF panels were ground flat to a thickness of 1.25 and 2.5 cm and each one paired with two 3-mm-thick 304 stainless steel plates. The sandwich panels are diffusion bonded in a vacuum furnace at 1200 °C. Samples for compression testing were cut using a Buehler Isomet 4000 diamond wafering blade, while the bending and tension samples were cut using a wire electric discharge machine (EDM). A section from the CMF-CSP was cut, ground, and polished using 320, 600, 800, 1200, and 2400 grit sandpaper for imaging and analysis of the interface between the face sheets and the CMF core. Digital imaging of the bonding surfaces was followed by scanning electron microscopy (SEM) using a Hitachi SU3500.

Table 1 Sample dimensions for the CMF-CSP and the base CMF samples (shown in parenthesis)

	Overall thickness (mm)	CMF thickness (mm)	Face sheet thickness (mm)	Width (mm)	Depth or length (mm)
Compression	31.20	24.70	3.25	25.4	25.4
Bending	17.50 (15.40)	11.10 (15.40)	3.2 (–)	38.04 (31.40)	110.05 (101.60)
Tension (Reduced area of dog-bone specimen)	17.51 (14.25)	11.16 (14.25)	3.18 (–)	12.70 (12.74)	50 (50)

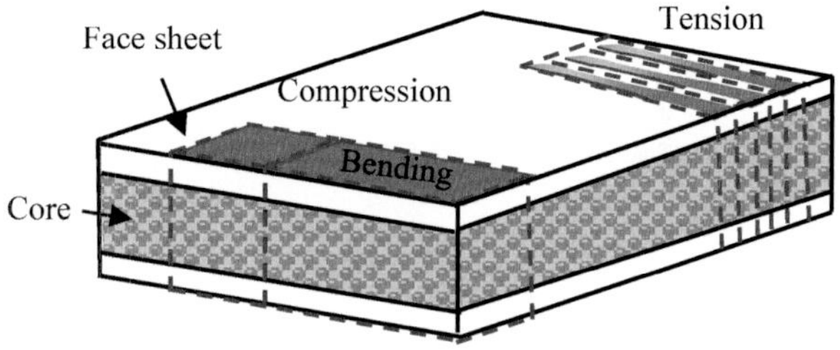

Fig. 1 The orientation where the samples were cut for testing. Note that this figure is for reference and is not to scale

Compression samples were cut to match ASTM E9 testing standard with a square cross section and a 1–1 height-to-depth ratio [6]. The bending sample size was modified following ASTM D7249, and the tension samples were cut to match the plate standard for ASTM E8 [7, 8]. The sizing of each sample, including their thickness and respective dimensions, is listed in Table 1 for the CMF-CSP and the base CMF samples used for comparison are shown in parentheses. The dimensions listed for the tensile tests are for the reduced area of the dog-bone specimens. Each sample was ground on the surfaces for clear imaging during testing using a 320 and 600 grit sandpaper. Figure 1 shows a drawing of the CMF-CSP and the orientation from which compression, bending, and tensile samples are cut from the panels. Multiple samples were tested for each mechanical test setup for validation.

Experimental Methods

Compression

Quasi-static compressive tests were run on a 980 kN load cell MTS 810 universal testing machine. The testing was conducted at a displacement rate of 1.27 mm/min. A digital camera was used to image each sample under loading. Polytetrafluoroethylene

(PTFE) sheets and graphite lubricant were placed at the interface between the sample and the platens to reduce friction and barreling effects during testing. Prior data of CMF under quasi-static compression was used for comparison.

Bending

The bending tests of CMF and CMF-CSP were conducted on a 98 kN load cell MTS QTest TM012 universal testing machine using a rate of 1 mm/min. All of the bending samples were tested with an inner span of 60 mm and an approximate overhang of 25 mm on each side. The three-point bending rollers were covered in 0.05-mm-thick PTFE sheet to reduce friction against the surface and allow for sliding during deformation and avoid pinning of the sample. The load and displacement are recorded by the computer through the frame's load cell and used to create the load–displacement curves for each test.

Tension

The tension samples were tested on the same MTS QTest TM012 universal testing machine with a 98 kN load cell. The tensile dog-bone samples were placed within pneumatic grips and were displaced at a rate of 1.27 mm/min. Spacers are used to ensure the samples are centered within the grips and loaded uniaxially. The load and displacement are recorded and used to create the stress–strain curves.

Results and Discussion

Microstructure and Interfacial Bond

Digital image of the CMF-CSP is shown in Fig. 2. As can be seen in this figure, the bond between the CMF surface and the stainless-steel face sheets is semi-continuous across the surface of the CMF. This is due to the open porosities that exist on the exposed CMF surface, mostly seen at the base of the image (some of these are marked by arrows). These spheres are opened along the surface during grinding to create a smooth surface to bond against the face sheets. SEM images of the surface contact between the CMF matrix and face sheets are shown in Fig. 3. From Fig. 3a, we can see a coherent bond between the CMF matrix and the stainless-steel face sheet (SS FS) (highlighted by the white arrows). Figure 3b gives a close up of the bonding at the sandwich panel interface between CMF matrix and steel face sheet. Figure 3c, d shows some locations with varying degrees of bonding and small gaps between

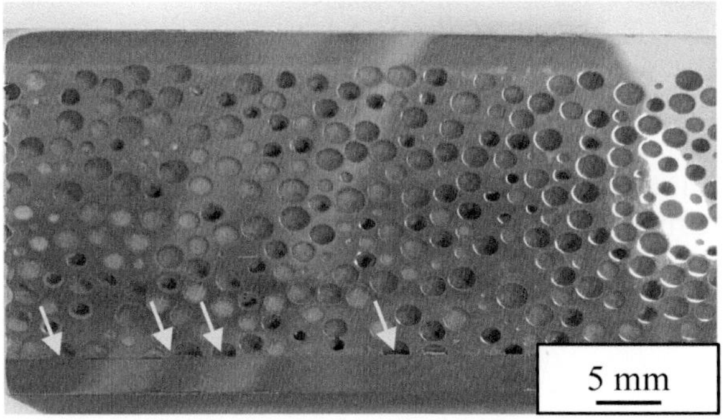

Fig. 2 Digital image of a CMF-CSP cross section with a 25 mm core and 3 mm face sheets on either side

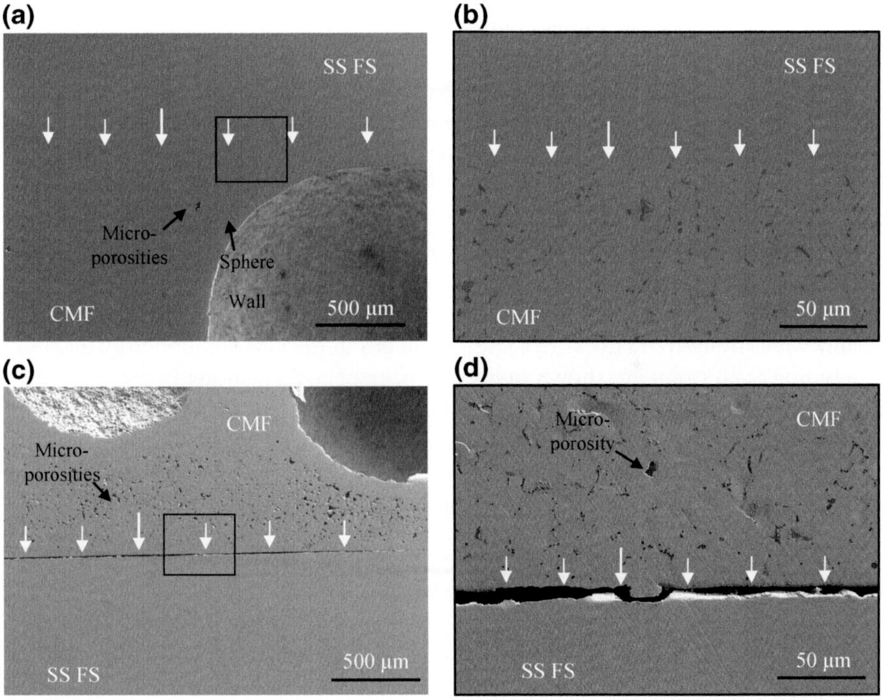

Fig. 3 SEM imaging of CMF-CSP and the interface between the CMF core and face sheets. (**a**) and (**b**) show a coherent bond between the CMF matrix and stainless steel face sheet. The white arrows highlight the interface line between the two bodies. Figure 3c, d shows an area in which a small gap exists between the CMF core and face sheet on the same sample

the CMF core and its face sheet. These gaps mostly appeared during secondary treatment of the sandwich panels indicating a poor bonding at those areas. The gap was measured to be approximately 20 μm at its widest point. The gap varies along the surface and completely disappears further along the sample length, giving the face sheet enough strength to remain bonded. Future work is expected to focus on optimizing and testing different methods of bonding to create the strongest product and avoid any gaps that can limit the strength of the sandwich panel.

Compression

The stress–strain curve of S-S CMF is presented in Fig. 4a alongside the stress–strain curve for the CMF-CSP in Fig. 4b. As can be seen in Fig. 4b, the multiple stress–strain curves of CMF-CSP under compression show consistency of the performance of the material. Moreover, the general shape of deformation is similar between the S-S CMF Fig. 4a and S-S CMF-CSP Fig. 4b curves. This is an indication that the face sheets on the sandwich panels transfer the load to the CMF core under compression and do not effectively contribute to load bearing under compression. The deviation between the previously tested CMF and the current tests most likely exist due to sample sizing. The current sets were cut from a plate with a total CMF thickness of 25 mm (height/width ratio of 1/1), limiting the overall sample size, whereas previous samples were manufactured in heights up to 90 mm (height/width ratio of about 4/1). The larger height-to-width ratio can yield a better representation of the total compressive strength of bulk CMF. Digital images of the CMF-CSP before and after undergoing 60% strain are shown in Fig. 5a, b, respectively, showing the uniformity of the deformation and complete elimination of its porosities. Further experiments of height/width ratio of 1/1 of CMF panels without face sheets will be conducted for comparison in the future. Potential grain growth and microstructural evolution of CMF during diffusion bonding is another possibility to affect the performance

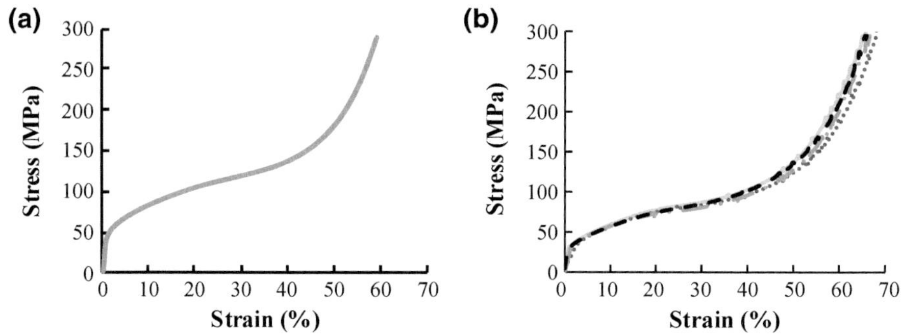

Fig. 4 Stress–strain curves of **a** 2 mm sphere S-S CMF and **b** four CMF-CSP under quasi-static compression. (**a**) Reproduced using data from [2]

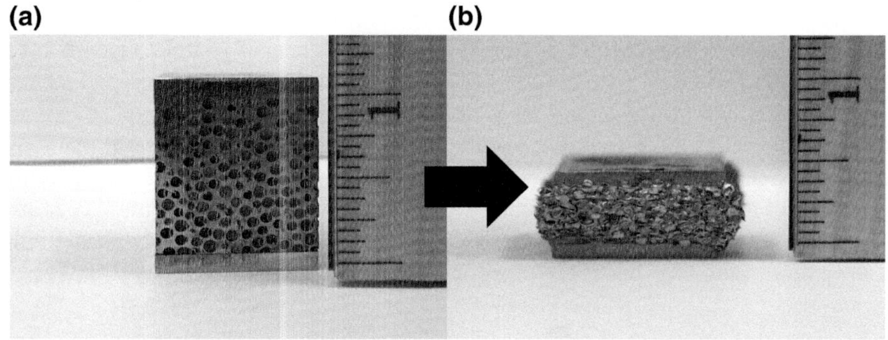

Fig. 5 Digital images of the CMF-CSP, **a** before and **b** after 60% compression

of CMF-CSP compared to bare CMF, and it is currently under investigation to be reported in future studies.

Bending

The flexural load–displacement plots for the bare CMF and CMF-CSP are plotted in Fig. 6a, b. The CMF samples yield at approximately 1.2 mm displacement, where the stress at the base of the sample exceeds its tensile strength and a crack begins to form. The maximum load is measured to be just below 12 kN for bare CMF. The addition of face sheets improves the failure to 5 mm displacement. The failure of the CMF core is preceded by three large peaks (Fig. 6b). These peaks are attributed to the shear failure between the face sheets and the CMF core at each of the pins in contact with the sample. The face sheets separate from the core and shift as the load

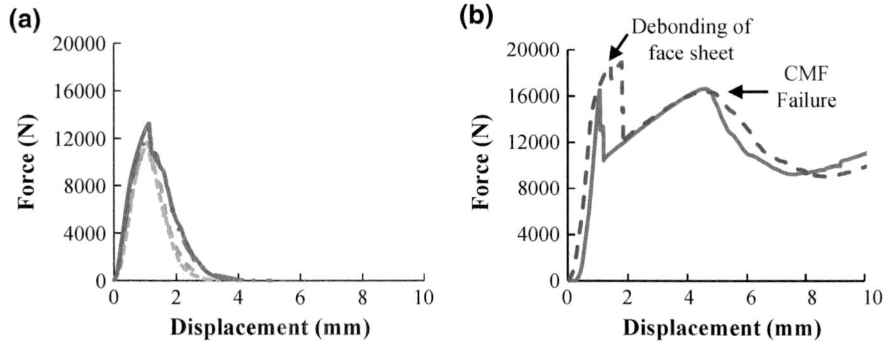

Fig. 6 Load–displacement curves of **a** four CMF samples and **b** two CMF-CSP under three-point bending

is transferred from the face sheets to the CMF core. At this point, the core begins to yield, similar to a pure CMF sample, reaching a peak stress of 16 kN. Once the core cracks, the stress is then transferred back to the face sheets and the bending reaches a plateau between 9 and 12 kN. It is noted that the face sheets helped to extend the core's life after the interfacial bonding fails between the layers. Other methods of adhering the two materials may strengthen the bond and reduce the number of failure peaks as well as possibly increase the stress at which they fail. Future tests are expected to explore the effect of face sheet thickness and optimize the bend strength of the overall sandwich panel while reducing its overall density.

Tension

The stress–strain curves of the tensile tests are plotted in Fig. 6. Metal foams are known to be weak under tension due to the inclusion of porosities and stress concentrations. However, due to the presence of a matrix between the porosities, S-S CMF normally offers better tensile strength compared to other types of metal foams. In this study, CMF only samples had an ultimate tensile strength of 80 MPa at approximately 7% strain (Fig. 7a). The CMF-CSPs show improved ductility and strength due to the presence of face sheets in Fig. 7b. The CMF core of the sandwich panels fractures earlier than the ductile stainless steel face sheets. The samples begin to crack at stresses as high as 210 MPa, almost 3 times higher than the bare CMF. Additionally, the life of the core is extended up to 28% strain before failure of the CMF core. Following the failure of the CMF core, the stress and strain values are calculated using the cross section of the face sheets only. The final strength of the sample aligns with the expected life of 304 stainless steel (face sheets).

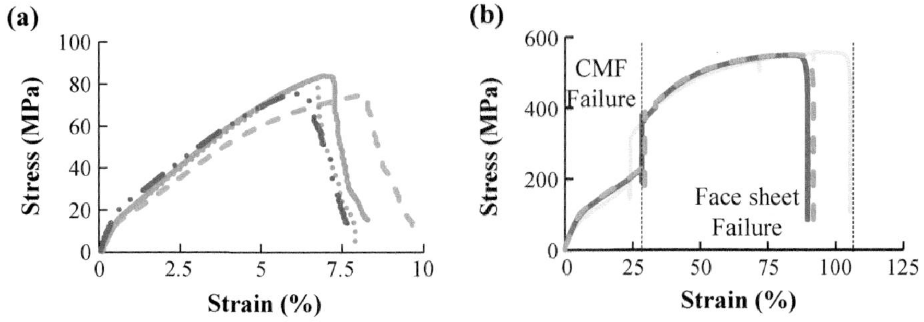

Fig. 7 Stress–strain curves of **a** four bare CMF samples and **b** three CMF-CSP under quasi-static tension. Figure 4b represents the CMF core failure on the left and the face sheet failure on the right

Conclusion

CMF core sandwich panels were manufactured using CMF and diffusion bonding them to stainless steel face sheets. Analyses of the interfacial bond between the CMF core and stainless-steel face sheets were carried out using SEM imaging. At the matrix/face sheet interface, the two bodies show good adhesion and form a coherent body. Small gaps were found in some areas, particularly with a high density of open sphere porosities. The mechanical properties of CMF and CMF-CSP were explored through compression, bending, and tension testing. The results of these three different load types are reported for quasi-static loading conditions. It can be seen that both the bending and tensile strengths of the CMF are improved by the addition of bulk metal face sheets. The compression of the CMF is not changed by the exterior layers, proving that it is able to retain its energy absorption capabilities. The bend life of the CMF-CSP was much longer than the bare CMF product due to the delay in fracture initiation and load spread through the core. A similar effect can be seen under tensile testing where, in addition to the sample's fracture strain, the fracture stress also improves by more than twofold. In addition to the added mechanical strength, CMF-CSP offer other benefits including a non-porous surface that can protect against corrosion. Future testing will analyze alternative bonding techniques and further optimization of the CMF-CSP under each loading condition.

Acknowledgements This research was supported by the Department of Transportation (DOT) Pipeline and Hazardous Materials Safety Administration (PHMSA) under project number DTPH5616C00001.

References

1. Ashby MF (2000) Metal foams: a design guide. Butterworth-Heinemann
2. Andrews E, Sanders W, Gibson LJ (1999) Compressive and tensile behavior of aluminum foams. Mater Sci Eng A 270(2):113–124
3. Neville BP, Rabiei A (2008) Composite metal foams processed through powder metallurgy. Mater Des. 29:388–396
4. Garcia-Avila M, Rabiei A (2015) Effect of sphere properties on microstructure and mechanical performance of cast composite metal foams. Metals 5:822–835. https://doi.org/10.3390/met5020822
5. Vendra LJ, Rabiei A (2007) A study on aluminum–steel composite metal foam processed by casting. Mater Sci Eng A 465:59–67
6. ASTM E9-19 (2019) Standard test methods of compression testing of metallic materials at room temperature. ASTM International, West Conshohocken, PA. www.astm.org
7. ASTM D7249/D7249M-18 (2018) Standard test method for face sheet properties of sandwich constructions by long beam flexure. ASTM International, West Conshohocken, PA. www.astm.org
8. ASTM E8/E8M-16a (2016) Standard test methods for tension testing of metallic materials. ASTM International, West Conshohocken, PA. www.astm.org

Part IV
Syntactic Foams

Design and Fabrication of Metal Matrix Syntactic Foams by Low-Pressure Injection Molding

M. Spratt, J. W. Newkirk and K. Chandrashekhara

Abstract Metal matrix syntactic foams are particulate composite foams composed of a supporting metal matrix and high-strength hollow particles. These materials have higher specific strengths and specific stiffnesses compared to solid metals and metal foams. The reinforced porosity in these foams is the primary cause of this increase in strength-to-density ratio. Metal matrix syntactic foams can be difficult to manufacture, however, without extensive fracture of the hollow particles. The first goal of this study was to optimize a water-based binder for low-pressure injection molding of metal matrix syntactic foams. The optimized binder composition was 7% agar, 4% glycerin, and 89% water. The second goal of this study was to test the material compatibility between silicate glass materials and copper alloys. It was found the common copper sintering aids reacted negatively with silicate glasses such that they melted at or below the sintering temperature of the metal. Pure copper does not have this issue, but it is conversely difficult to sinter to full density.

Keywords Syntactic foam · Injection molding · Metal

Introduction

Syntactic foams are versatile composites with many potential applications. These materials are comprised of a matrix material and hollow particles embedded in that matrix. This classifies them as both a particulate composite and a foam. Because the porosity in the matrix is reinforced, these materials exhibit excellent specific strengths and stiffnesses. They function well in applications where lightweight, structural materials are needed [1]. For example, polymer-glass syntactic foams have been used as deep-sea diving vessels. Hollow cenospheres (a coal by-product) added to ceramic bricks both improve thermal insulation and lower density. Metal matrix syntactic

M. Spratt · J. W. Newkirk (✉)
Materials Science Department, Missouri University of Science and Technology, Rolla, USA
e-mail: jnewkirk@mst.edu

K. Chandrashekhara
Mechanical Engineering Department, Missouri University of Science and Technology, Rolla, USA

N. Dukhan (ed.), *Proceedings of the 11th International Conference on Porous Metals and Metallic Foams (MetFoam 2019)*, The Minerals, Metals & Materials Series, https://doi.org/10.1007/978-3-030-42798-6_9

foams foams (MMSFs) have not yet been used commercially to any great extent. There has been significant research effort, however, to produce quality MMSFs [2]. Work done by Garcia-Avila et al. [3] in making lightweight MMSF armor shows some of the military interest in these materials. Potential commercial applications include transportation (aerospace and automotive) and packaging [2].

MMSF materials function as both metal–ceramic composites and metal foams. The particles, even though they are hollow shells, do act as load-bearing members of the composite, as shown by Balch et al. [4]. Sudarshan and Surappa [5] showed that they also contribute to the composite's wear resistance. The porosity is of a specific size range determined by the size of the ceramic particles and the cell wall thickness. This closed porosity will absorb energy [6] and also contribute to thermal insulation. The combination of hollow spheres and matrix can dampen out vibrations—acoustic and otherwise—as well [1, 7]. As in composites, the properties of the MMSF can be tailored to specific applications by selecting appropriate materials.

Quality MMSFs are difficult to manufacture. The thin-walled hollow particles and the high-temperature interactions between the composite materials can cause the majority of the issues. The hollow particle cell wall size usually depends on the size of the particles. The shells can fracture relatively easily during processing. This is especially true when strong shear forces are applied to the particles, as in stir casting. The particles also generally have very low densities and will float in molten metal. For powder metallurgy processes, the hollow particles can be difficult or impossible to dry mix into metal powder. These issues limit both the molten metal processes and powder metallurgy processes that can be used effectively. When heated, the metal matrix can react with the ceramic or glass hollow particles in several ways. Diffusion will occur between the two materials. Compounds and precipitates can form. The introduced elements can also alloy with the metal. Any or all of these possibilities can be detrimental to the composite microstructure and properties. The melting or softening points of the ceramic or glass material must also be significantly higher than the melting or sintering temperature of the metal. This ensures that the hollow particles will not melt or slump during processing.

There are two processing avenues typically used in literature to manufacture MMSFs. First, and most commonly, are molten metal processes. In these methods, molten metal is combined with the ceramic particles and then formed into a part. Stir casting and pressure infiltration are two examples of this method [1]. This is usually a fast operation—the materials are not typically in contact for long. Particle flotation can be an issue, however. Some methods of combining the metal and ceramic particles, such as stirring with an impellor, can damage the hollow particles. Second, a powder metallurgy process can be used. In these methods, powders are combined, formed, and sintered. Injection molding and powder pressing are two examples of this method. This process requires the powders to be in contact at high temperatures for significant periods.

The goal of this work was twofold. The first goal was the creation of a generalized procedure for low-pressure injection molding of metal matrix syntactic foams. For this goal, a water-based binder was selected and optimized. The second goal was to test the compatibility between copper alloys and glass materials to identify potentially viable MMSF systems.

Procedure

Binder Optimization

The agar–glycerin–water binder used in this experiment was developed for this system using a mixture model. Because there were three components, a three-component simplex design was used. A cubic response model was used to optimize the binder for green and sintered density after collecting the data. Ten sample compositions are used in this model. Each binder composition was 45 vol.% of the final specimen with the bronze powder at 33 vol.% and glass microspheres filling the rest of the space. The response variables measured were the green, dried, and sintered density. Three replications were made for each sample. Each of the 30 specimens was made and measured in a randomized order. JMP®, a statistics program, was used to design this experiment and calculate the response surfaces. A confidence of 95% was used for this study. The fitness of the model was calculated using the F Statistic. If the calculated F-ratio (Eq. 1) from the experimental data was less than the tabulated F-ratio, the model was considered to be a reasonable representation of the response surface of the mixture. This was the case for each of the measured response variables tested in this experiment.

$$F = \frac{n * s_{\bar{x}}^2}{s_{\text{pooled}}^2} \quad (1)$$

where n is the sample size, $s_{\bar{x}}^2$ is the variance of the sample means, and s_{pooled}^2 is the mean of the sample variances.

The model's lower limits were selected based on the composition in German's book [8]. The lower bound of the agar was 4 vol.%, the glycerin was 0 vol.%, and the water was 85 vol.%. The compositions for the test samples were determined using these lower bounds. The ten binder compositions tested in this experiment are listed in Table 1.

Material Compatibility Study

Combinations of different materials were made to test material compatibility. Table 2 shows each material with the supplier, chemistry, particle size data provided by the supplier, and the morphology.

Table 1 Binder compositions for each sample in the three-component mixture model

	Agar (%)	Glycerin (%)	Water (%)
1	15.0	0.0	85.0
2	4.0	11.0	85.0
3	4.0	0.0	96.0
4	9.5	5.5	85.0
5	9.5	0.0	90.5
6	4.0	5.5	90.5
7	7.7	3.7	88.7
8	11.3	1.8	86.8
9	5.8	7.3	86.8
10	5.8	1.8	92.3

Table 2 Each material used in the study is listed

Material	Supplier	Chemistry	Particle size data (mesh size %)	Density (g/cc)	Morphology
Borosilicate glass	MO-SCI Corp.	70–85% SiO_2 10–15% B_2O_3 5–10% Na_2O 2–5% Al_2O_3	−140	0.15	Hollow, spherical
Silica	The ceramic shop	SiO_2	−325	2.65	Solid, angular
Copper	Royal Metal Powders Inc.,	99.8% Copper 0.06% Hydrogen loss	+325 0.4% −325 99.6%	8.94	Solid, spherical
Bronze	Royal Metal Powders Inc.,	88.46% Copper 11.3% Tin 0.24% Phosphorous	+100 0.1% +140 20.8% +200 5.5% +325 21.6% −325 4.0%	8.73	Solid, spherical
Brass	ARTMOLDS	Cu_3Zn_2	−325	8.73	Solid, spherical

For each material combination, the composition of the composite was 60 vol.% metal powder. Except for the copper–borosilicate glass combination, each combination of metal and ceramic powder was made. Each sample consisted of three specimens, made in a random order as described in the sample preparation section of the procedure. Bronze and brass samples were sintered at 900 °C for 1 h. The copper sample was sintered for 1000 °C for 3 h.

Sample Preparation

The binder was prepared by first heating de-ionized water on a hot plate to boiling. The agar and glycerin were then added. The solution was agitated for approximately 5 min to form a stable gel. The metal and ceramic powders were heated to between 50 and 100 °C before adding them to the binder.

During binder optimization and material compatibility testing, the samples were cast into silicone molds. The powders were added to the binder on the hot plate and vigorously stirred for at least five minutes. The solution was poured into the silicone molds and allowed to cool before demolding.

Thirty pure copper specimens were injection molded in this work using the optimized binder and a solids loading of about 60 vol.%. The injection molding machine used in this experiment was a Peltsman MIGL-28. This machine has a reservoir that holds the IM feedstock. The reservoir includes a paddle that turns at a fixed rate of 58 rpm. The working temperature of this binder is between 50 and 100 °C. Typically, the temperature was kept at 80 °C. To prevent seizing, the binder was added to the heated reservoir first. Then, the hot powders were added slowly while the paddles stirred the solution. After adding the powders, binder additions of about 2% were made as needed. Minor adjustments to viscosity were done by altering the temperature by 10 K. The slurry was mixed for one hour after the last binder addition. Ten psi of gas pressure, held for 10 s, was used to push the slurry into the mold. The mold used in this experiment was a steel rectangular prism with the following dimensions: $60 \times 9.9 \times 6.9$ mm^3. The specimens were demolded after 30 s.

Sample Densification and Measurement

Samples were dried for at least 12 h at 120 °C to de-water the part. They were debound at 450 °C for 1 h in flowing air. The samples were then held at 450 °C for 1 h in flowing argon. Samples were sintered directly from debinding at a temperature depending on the alloy.

To measure the green density and shrinkage, the length, width, and height of each specimen were measured with calipers. Sintered density was measured by Archimedes' method as described in ASTM C373. The boiling method was used to impregnate the specimens with water. Geometric density was also measured with calipers. An ASPEX machine was used to do automated inclusion size and chemistry analysis of specimen microstructure. An ASPEX machine is a SEM with an attached EDS detector that can use software to automatically detect the features of a micrograph.

Results and Discussion

Binder Optimization

For this study, a binder with several requirements beyond those typical for most injection molding was necessary. First, it had to have a working temperature below 120 °C. This was a machine limitation. Many common polymer binders, which only become fluid enough for injection molding above 120 °C, were not suitable. Second, there could be no feedstock compounding and pulverizing step. One method of making injection molding feedstock is to compound the binder (often a thermoplastic) with the powders, then either pelletizing or pulverizing the feedstock for injection molding. The hollow particles are too delicate to submit to compounding or pulverization. Finally, a binder that was non-toxic, simple to clean up, and did not require harsh chemicals for debinding would be ideal. The initial binder composition for this experiment was selected from German [8]. This binder uses agar, a substance similar to gelatin, to form a gel with water. Glycerin is used to support the polymer chains. This binder requires a drying step, which removes the majority of the water. It then requires a debinding step, which occurs at around 400 °C according to Porter [9]. It also fits all the criteria listed.

The results for the binder optimization are shown in Fig. 1. Figure 1a shows the calculated green density of the samples after the samples were fabricated. The highest density (3.96 g/cc) occurred at 6.6% agar, 3.9% glycerin, and 89.6% water. Figure 1b shows the calculated dry density of each sample after 24 h in a drying oven at 120 °C. The highest density (3.62 g/cc) occurred at 6.5% agar, 3.9% glycerin, and 89.7% water. Figure 1c shows the calculated sintered density of each sample after sintering at 900 °C for 1 h. The highest density (4.86 g/cc) occurred at 6.8% agar, 3.9% glycerin, and 89.4% water. The optimized binder was taken to be 7% agar, 4% glycerin, and 89% water.

The density of the specimens decreased after drying. Because water expands as it heats up, swelling occurred which caused the volume of the specimens to increase slightly during the drying step. Figure 2 shows a graph of the shrinkage (negative percent shrinkage indicates swelling) of each sample after 1, 2, 3, and 24 h of drying. Shrinkage, in this case, refers to the change in length of the samples over the initial length.

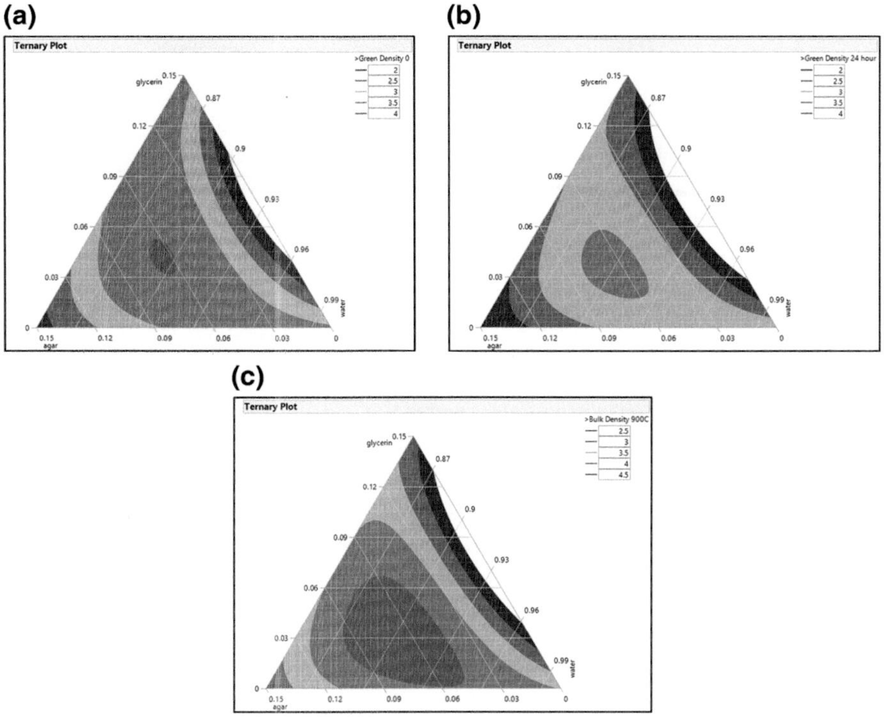

Fig. 1 The response surfaces for the green density **a** the dry density **b** and the sintered density. **c** Each color of the graph shows where the density was greater or equal to the legend

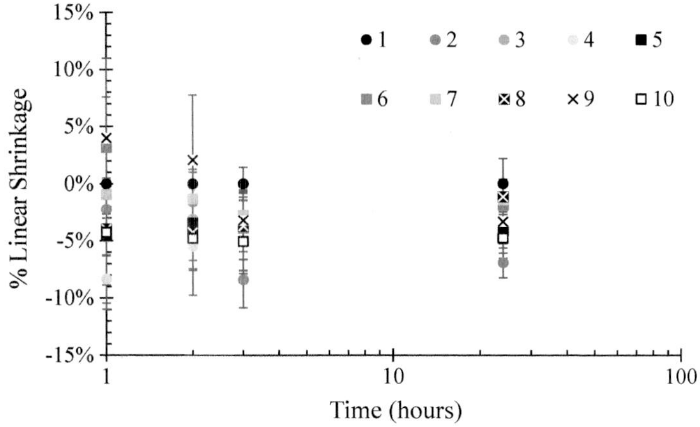

Fig. 2 Linear shrinkage of samples over 24 h. Negative values indicate swelling

Material Compatibility

Several material systems were explored in this work. The matrix materials tested were copper, brass, and bronze. Bronze and brass both have relatively low melting temperatures (about 950 °C) and high density. The low melting temperature, and therefore low sintering temperature, was beneficial as it allows for lower temperature glass materials to be used. The differences between the density of the copper alloy metals and the hollow borosilicate glass were about as extreme as any metal matrix syntactic foam is likely to be. This was useful when creating a generalized procedure for MMSFs, as this study intended.

Bronze and brass are both copper alloys with different alloying elements. In brass, zinc is the primary alloying element. In bronze, tin is the primary alloying element. Both of the alloying elements lowered the melting temperature of the copper. The matrix was densified by pressureless sintering at or above 90% of the melting temperature of the metal. When combined with silicate glass materials, both alloying elements (tin and zinc) diffused out of the copper and into the silicate glass. Tin oxide and zinc oxide also formed. This left the brass and bronze particles depleted of their alloying elements, which raised the sintering temperature of the alloys. Figure 3 shows an SEM micrograph of bronze powder with 40 vol.% solid silica. Phases on this figure were identified using SEM-EDS. The solid silica particles, the gray phase, appeared to have completely lost their shape and, essentially, liquid-phase sintered the part. The bronze, on the other hand, remained spherical. The melting temperature of pure silica is well above the 900 °C sintering temperature used here, indicating that the diffusion of the tin into the silica lowered the melting temperature. A similar response was seen in the brass–silica system.

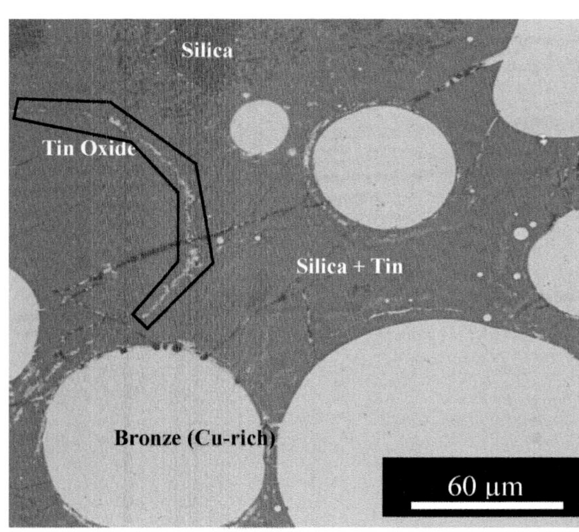

Fig. 3 Phase identification in 40 vol.% solid silica, 60 vol.% bronze specimens

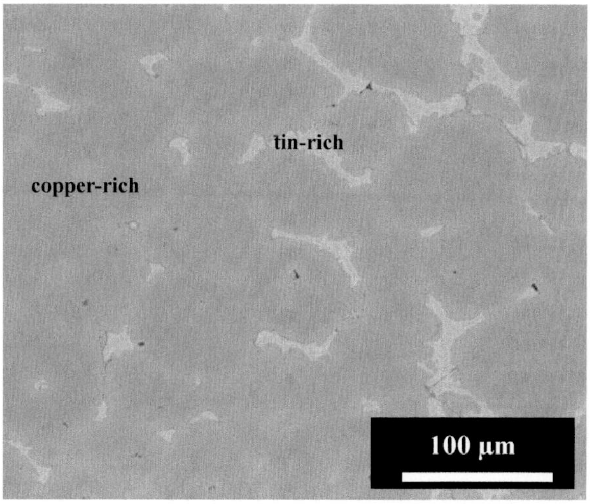

Fig. 4 Bronze with 40 vol.% hollow borosilicate glass after sintering. No sign of the glass after sintering

Figure 4 shows the bronze with 40 vol.% borosilicate bubbles after sintering. The lighter gray phase is a tin-rich phase, and the dark phase is copper-rich. In this image, the bronze was able to sinter together adequately. The borosilicate hollow spheres, due to their low softening temperature (around 650 °C) and low weight in the system (<2 wt%), were not found in the structure. The borosilicate bubbles may have collapsed due to the high sintering temperatures. It is also possible that they were diffused into the bronze structure entirely. Again, this was repeated with the brass material.

In an attempt to avoid the unwanted reactions between the alloying elements and the ceramic materials, pure copper powder was used. Copper has a melting temperature around 1050 °C, which eliminated the hollow borosilicate glass as a viable candidate. Figure 5 shows one specimen made with 15 vol.% silica and copper. The silica (the dark phase) and copper (the light phase) are both visible in the micrograph. The angular silica particles retained their shape after sintering. Also present is a significant amount of porosity, a consequence of not having a sintering aid present. An inclusion analysis of this specimen was performed via ASPEX SEM. Some of the "inclusions" analyzed, particularly the high-copper inclusions, were likely porosity. Figure 6 shows the results of the inclusion analysis, where over 900 inclusions were counted. Most of the clustering appears around the Si–O line of the ternary diagram. Copper diffusion into the silica did appear to be significant, however, with the major cluster stretching out to about 30% copper. The copper–silica system appeared to work reasonably well for this work.

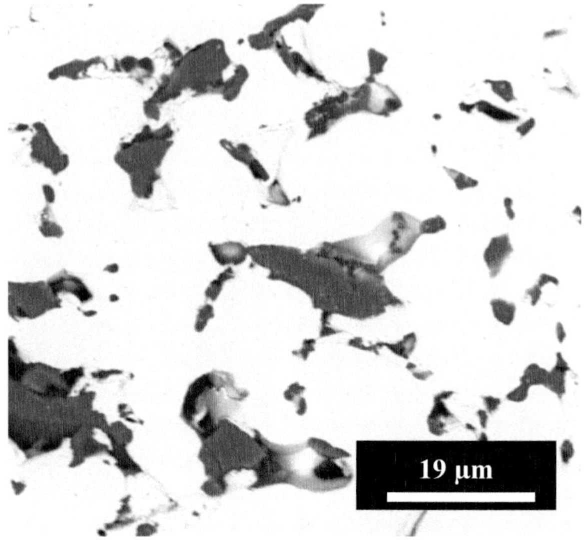

Fig. 5 The copper and 15 vol.% solid silica SEM micrograph after sintering

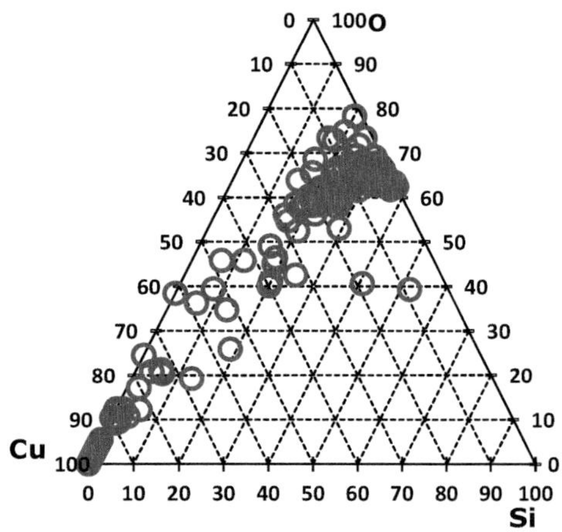

Fig. 6 Inclusion analysis results from silica inclusions in copper after sintering

Injection Molding

Pure copper specimens were injection molded to develop the injection molding procedure. They were used to adjust the sintering time to decrease porosity. Sintering was carried out at 1000 °C for several times. Figure 7 shows the average porosity of each sample after each sintering time. These values are averages of 30 specimens. As expected, longer sintering times resulted in less porosity. The highest density attained was 7.82 ± 0.04 g/cc at the 10-h sintering time. The sample, after 10 h, still retained 10% closed porosity and 3% open porosity. The samples were densified from 40% open porosity without pressure or a sintering aid. For optimal densities

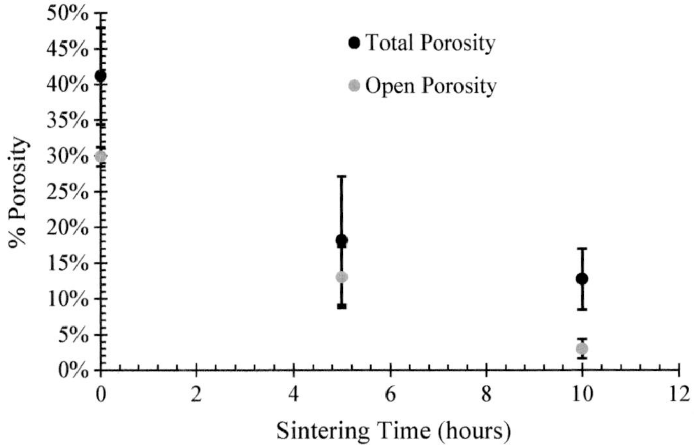

Fig. 7 Total and closed porosity in copper samples after sintering. A theoretical density 8.96 g/cc was assumed

above 99% dense, a sintering aid or added pressure is likely necessary. However, as was shown, the sintering aid is required to be one that does not react negatively with the ceramic material.

Glass materials are among the most common form of hollow spheres available. This is especially true for size ranges appropriate for powder metallurgy techniques. Thus, many researchers are either forgoing powder metallurgy processes or using glass hollow spheres. Unfortunately, glass materials are known for their open, non-crystalline atomic structures, which allow for the easy incorporation of many different elements into the material. This can change the melting temperature or cause crystalline phases to form. The melting temperature change is often most likely, as any sintering aid is likely to be a low melting temperature material. This can cause the glass to melt prematurely or slump during sintering. At that point, the material is no longer a syntactic foam. Refractory ceramic materials, such as alumina and silicon carbide, would potentially be better alternatives for this system.

It is necessary to develop new ways of manufacturing refractory ceramic hollow particles in micron-sized ranges. Currently, these materials are available in millimeter-sized ranges. It is rare to find quality micron-sized hollow powder that is not a glass. This is important because most powder metallurgy techniques rely on micron-sized powders. An increase in the material availability of ceramic powders would not only increase the tailor-ability of MMSFs, but also the processing landscape. Additive manufacturing, injection molding, powder pressing, etc. would be potentially viable MMSF manufacturing options.

Conclusions

A water-based gel binder was optimized for injection molding of metal matrix syntactic foam materials. The composition that proved to have the highest density results was 7% agar, 4% glycerin, and 89% water. Material compatibility was tested between three copper alloys and silica-based glasses. The alloying elements in the copper alloys reacted negatively with the silica glasses to lower the melting temperature of the glass. This caused the glass to melt in the matrix, which is not acceptable for syntactic foam processing. Pure copper powder did not react negatively with silica glass during sintering. However, even after sintering at 93% of the copper melting temperature for 10 h, the total porosity was $13 \pm 3\%$. To use a copper alloy that is easier to sinter to full density, a micron-sized hollow refractory ceramic material is recommended.

References

1. Rohatgi PK, Gupta N, Schultz BF, Luong DD (2011) The synthesis, compressive properties, and applications of metal matrix syntactic foams. JOM 63:36–42
2. Gupta N, Rohatgi PK (2014) Metal matrix syntactic foams: processing, microstructure, properties and applications. DEStech Publications, Inc., https://books.google.com/books/about/Metal_Matrix_Syntactic_Foams.html?id=yfjuAwAAQBAJ&pgis=1, Accessed 16 Oct 2015
3. Garcia-Avila M, Portanova M, Rabiei A (2015) Ballistic performance of composite metal foams. Compos Struct 125:202–211. https://doi.org/10.1016/j.compstruct.2015.01.031
4. Balch DK, Dunand DC (2006) Load partitioning in aluminum syntactic foams containing ceramic microspheres. Acta Mater 54:1501–1511
5. Sudarshan MK, Surappa (2008) Dry sliding wear of fly ash particle reinforced A356 Al composites. Wear 365:349–360
6. Rabiei A, Vendra LJ (2009) A comparison of composite metal foam's properties and other comparable metal foams. Mater Lett 63:533–536. https://doi.org/10.1016/j.matlet.2008.11.002
7. Zhang J, Perez RJ, Lavernia EJ (1994) Effect of SiC and graphite particulates on the damping behavior of metal matrix composites. Acta Metall Mater 42:395–409. https://doi.org/10.1016/0956-7151(94)90495-2
8. German RM, Bose A (1997) Injection molding of metals and ceramics. Metal powder industries federation
9. Porter M-A (2003) Effects of binder systems for metal injection moulding. http://www.diva-portal.org/smash/get/diva2:1017275/FULLTEXT01.pdf, Accessed 9 July 2018

The Deformation of Expanded Clay Syntactic Foams During Compression Characterized by Acoustic Emission

Csilla Kádár, Kristián Máthis, František Chmelík, Michal Knapek and Imre N. Orbulov

Abstract The deformation and failure mechanisms in syntactic foams with different metal matrices were investigated in this study. The syntactic foams were produced by the infiltration method using lightweight expanded clay particles (LECA) as a space holder and commercially pure Al or eutectic Al-12%Si alloy for the matrices. The samples were compressed at room temperature; simultaneously, the acoustic emission (AE) response and the surface deformation were monitored and collated with the deformation curves. In the case of the Al foam, ductile behavior with long plateau stress was observed. During this plateau regime, multiple active shear bands were detected. In contrast, the AlSi12 foam exhibited more brittle deformation behavior. At the end of the quasi-linear stage, the localization of the strain started within one large shear band, accompanied by a significant stress drop. The AE analysis revealed that the deformation was mostly governed by the fracture of LECA particles and the plastic deformation of the matrix material for both types of foams. After strain localization, in the case of the AlSi12 foam, the fracture of the matrix became significant, causing the deterioration of the sample. In Al foam, due to multiple active shear bands, the stress–strain curve is reproducible even for samples with four times of the size of the LECA particle.

Keywords Metal matrix syntactic foams · Deformation · Acoustic emission

---

C. Kádár (✉) · I. N. Orbulov
Department of Materials Science and Engineering, Budapest University of Technology and Economics, Műegyetem rakpart 3., Budapest 1111, Hungary
e-mail: kadar@metal.elte.hu

MTA–BME Lendület Composite Metal Foams Research Group, Műegyetem rakpart 3., Budapest 1111, Hungary

K. Máthis · F. Chmelík · M. Knapek
Department of Physics of Materials, Charles University, Ke Karlovu 5, CZ12116 Prague 2, Prague, Czech Republic

M. Knapek
Nuclear Physics Institute, Czech Academy of Sciences, Řež 130, 250 68 Řež, Prague, Czech Republic

N. Dukhan (ed.), *Proceedings of the 11th International Conference on Porous Metals and Metallic Foams (MetFoam 2019)*, The Minerals, Metals & Materials Series, https://doi.org/10.1007/978-3-030-42798-6_10

Introduction

In recent years, research on the mechanical properties of metal matrix foams and metal matrix syntactic foams (MMSF) has received extensive attention since these materials are lightweight structures with high strength-to-weight and stiffness-to-weight ratios, good damping and energy absorption properties [1–3].

There are different strategies for improving the strength and the energy absorption capacity of a foam. One option is to increase the strength of the matrix material. This can be achieved by (a) thermal or thermomechanical treatment, (b) adding alloying elements to form solid solution or to generate intermetallic phases, or (c) using micro- or nano-sized reinforcing particles such as SiC or alumina particles or carbon nanotubes [2]. However, there is another way to enhance the mechanical properties of a foam by incorporating low-density, stiff and porous particulates (filler material) into the matrix. The materials produced in this way are referred to as syntactic foams [4].

The most frequently used additions for manufacturing syntactic foams are the hollow spheres. Mostly, ceramic hollow spheres are used [5–7]; however, recently, MMSF with steel hollow spheres has also been produced [8]. Although hollow spheres enhance the plateau stress of the foam considerably, due to the high cost of producing new low-cost filler materials are needed. Up to now, syntactic foams with expanded glass [9], pumice [10], expanded perlite [1], and fly ash [4] fillings have been investigated.

Recently, lightweight expanded clay (LECA) as a filler material has also been designed as it features high-porosity, low-density, and good thermal/acoustic insulation properties [11, 12]. So far, only the effect of the LECA particle diameter has been investigated [12]. Puga and co-workers have found that with increasing LECA particle diameter both the yield strength and the densification strain increase. They found a change in deformation mechanism: for foams containing LECA particles larger than 7 mm, the deformation is less uniform [12].

In this paper, the effect of the matrix material on the failure mechanisms and compressive strength of MMSF containing LECA particles was studied. The changes in deformation mechanisms were also investigated by the acoustic emission (AE) and digital image correlation (DIC) techniques.

Experimental Procedure

The matrix materials of the investigated syntactic foams are a commercially pure aluminum (Al: 99.5 wt% Al, 0.2 wt% Si, 0.3 wt% Fe) and an AlSi12 alloy (86 wt% Al, 12.8 wt% Si, 0.1 wt% Fe, 1.1 wt% other). The syntactic foams were produced by low-pressure infiltration, as follows. In a closed profile crucible of a cross-section of 50 mm × 50 mm, first sieved LECA particles (chemical composition: 49 wt% O, 23 wt% Si, 13 wt% Al, 7 wt% Fe, 8 wt% others) having a diameter of 6–7 mm, and then a stainless steel grid (to hold them in position during infiltration) and a block of metal (AlSi12/Al) were placed. Next, the crucible (closed by welding) was heated up and held at 750 °C for Al and 650 °C for AlSi12 for 30 min. To complete the infiltration, inert gas (He) was applied for 120 s at 1 bar pressure. Subsequently, the block was quenched into water of ambient temperature. Finally, 4 mm of the side parts and the bottom of the obtained foam block were mechanically removed to ensure a homogeneous foam structure.

An MTS810 type servo-hydraulic universal testing machine (MTS Systems Corporation, Eden Prairie, MN, USA) was used to carry out the uniaxial compression tests at room temperature. Three cube-shaped foam specimens of each type with a side length of 30 mm and a density of 1.55 ± 0.05 g/cm^3 were compressed at a crosshead speed of 0.01 mm/s. Teflon strips were applied between the foams and machine grips to reduce interface friction. The deformation of the surface was recorded by a video camera. The DIC technique was applied to determine the strain localization during deformation by using the *ncorr* script implemented in Matlab [13].

Simultaneously with the compression test, the AE activity was monitored by the Vallen AMSY-6 System operating in a transient recording mode with a sampling rate of 1 MHz. The AE signal recorded using the Physical Acoustic Corporation WSα sensor was pre-amplified with a gain of 40 dB. The sensor was mounted on the surface of the specimen using vacuum grease and an elastic band.

The structures of the syntactic foams and the LECA particles were investigated by a Hitachi TM4000 Tabletop scanning electron microscope (SEM).

Results and Discussion

Representative engineering stress–strain curves with different matrix materials are shown in Fig. 1. In the case of unalloyed Al-LECA foam, ductile behavior can be observed (Fig. 1a), i.e., after the quasi-linear region the stress reaches a plateau, after which there is practically no stress drop. In the plateau region, the deformation of the cells in the shear band(s) takes place. Finally, in the densification stage, the matrix material fills the remaining pores produced by the crumbling of the LECA particles. Contrary to the Al foam, the AlSi12-LECA foam does not exhibit distinct plateau stress, but rather peak stress followed by a stress drop occurs. The peak stress of the AlSi12 foam is roughly twice as large as the peak stress of the Al foam. It is worth noting that even though the samples are quite small compared to the size of the LECA particles, the stress–strain response is reproducible for Al-LECA foams (see Fig. 1a), while for AlSi12-LECA the shape of the deformation curve depends on the structure of the foam specimen. Nevertheless, the variation of the peak stress was less than 10%.

The AE response during compression is dependent on the matrix material. The AE response shows a maximum around the peak stress in the case of Al-LECA foam, while for the AlSi-LECA foam the maximum AE activity can be measured before the peak stress is reached. In both cases, the AE peak is followed by a considerable

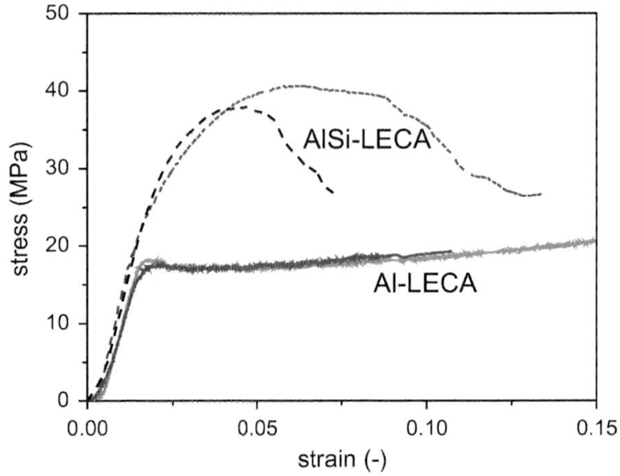

Fig. 1 The stress–strain curves for metal matrix syntactic (MMS) foams containing expanded clay particles with Al matrix (solid) and AlSi12 matrix (dash)

drop in the AE activity, especially for the AlSi12-LECA foam (Fig. 2). This drop is due to the localization of deformation. Due to strain localization, the number of active AE sources decreases and, correspondingly, the AE activity reduces.

The adaptive sequential k-means (ASK) clustering analysis was performed on the AE data to determine the dominant deformation mechanism as a function of strain. The AE data are sectioned into time windows (in this evaluation 1 ms time window was used) and the statistical parameters of the power spectral density function (such as mean frequency, peak amplitude, energy, kurtosis, etc.) of each time window are determined and grouped into clusters according to the algorithm of Pomponi and Vinogradov. The detailed description of the method can be found in [14]. For the identification of the AE sources, complementary methods like DIC and SEM were used. It is important to emphasize that if at a given strain multiple different AE

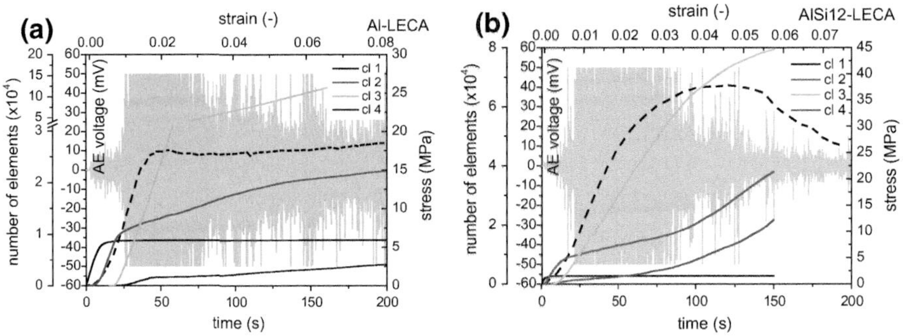

Fig. 2 Representative compressive stress–strain curves (dashed line), AE signals (gray line) and the time evolution of the cumulative number of elements in different AE clusters of **a** Al foam; **b** AlSi12 foam. Black line—noise; red line—plastic deformation of matrix; green line—fracture of LECA particles; blue line—fracture of the matrix

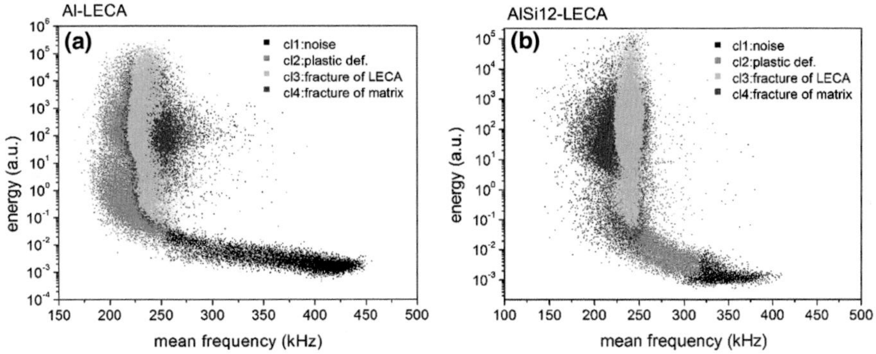

Fig. 3 The 2D projection of the clusters to the energy-mean frequency space for **a** Al foam; **b** AlSi12 foam

sources are active, the ASK method is only capable of determining the dominant one.

At the beginning of the test, the cluster of background noise should be identified. To that end, the AE measurement is started prior to the compression. During the test, the ASK algorithm identified four clusters for both, Al and AlSi12 foams. Figure 2 shows the time evolution of the cumulative number of elements for the different AE clusters. The following source mechanisms were assigned to the clusters:

Cluster 1: Noise (color code in figures: black)
The cluster appearing first is always the noise cluster, since the AE test starts prior to the deformation and, therefore, the noise cluster is not directly related to the deformation processes. This cluster contains low-energy AE signals with low amplitude and no sharp peaks (see also Fig. 3).

Cluster 2: Plastic deformation of the matrix (color code in figures: red)
In both cases, the second cluster appears at very low strain. Similarly to the analysis in our previous work on Al-based MMC containing hollow ceramic spheres [15] (e.g., the "drop-like" shape of the cluster (Fig. 3), the shape of the waveform), plastic deformation can be assigned to Cluster 2.

Cluster 3: Fracture of LECA particles (color code in figures: green)
The typical signal belonging to this cluster has a burst-like character with very short rise time and high amplitude, as observed also in earlier studies on MMSFs [15]. Furthermore, the energy range of these signals is wide, suggesting that the origin of AE is the fracture of LECA cell walls of different thicknesses. The cracking of the LECA cell walls at low stress is visible in the post-mortem SEM pictures (Fig. 4a).

Cluster 4: Fracture of the matrix (color code in figures: blue)
As for the AE signals in Cluster 3, the signals in Cluster 4 have wide frequency and energy ranges and negligible rise time, as observed in [15]. The video recording shows that in the case of the AlSi12 foam the role of matrix fracture becomes significant after strain localization (Fig. 4c). Figure 2b shows that the cumulative number of

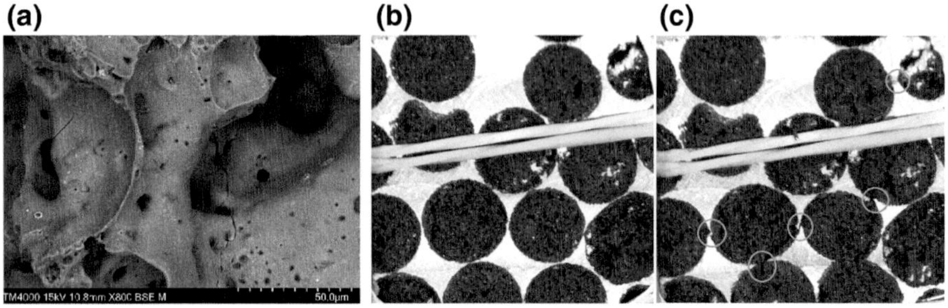

Fig. 4 The deformation of the reinforcement and the foam. **a** Cracks in the LECA particle in the deformed foam; **b** undeformed AlSi12 LECA foam, and **c** AlSi12 LECA foam at 0.06 strain. (The places where plastic deformation or fracture of the matrix took place are marked by red circles.)

elements in Cluster 4 suddenly increases when the AE activity decreases. It is indeed expected that due to strain localization the fracture of the matrix becomes more dominant (Fig. 4c). In the case of the Al foam, the fracture of thin cell walls between LECA particles is visible after strain localization.

The results of the ASK analysis show that the Al-LECA foam first deforms by plastic deformation of the matrix material. Due to the low strength of the LECA particles, the cracking of the LECA spheres starts at low applied stress (~5 MPa). The localization of the strain starts at the end of the quasi-linear stage (Fig. 5), and the formation of deformation bands can be subsequently observed. Contrary to conventional foams, the deformation of the Al-LECA foam is not restricted to only one band (Fig. 5). As a result, the AE activity decreases due to strain localization, but the drop in AE activity is only moderate since the deformation still takes place in a relatively large volume. After the appearance of deformation bands, the rate of growth in the number of elements in Cluster 2 decreases, while the number of elements in Cluster 3 increases linearly as a function of strain. It implies that the dominant deformation mechanism is the fracture of LECA particles throughout the deformation process.

For small strains, the AlSi12 foam deforms similarly to the Al foam, i.e., plastic deformation of the matrix takes place almost as soon as the compression starts, followed by the beginning of fracture of the cell walls of LECA particles at about 2 MPa applied stress. The onset of the LECA particles cracking at lower stress is due to the higher back stress of the AlSi12 foam compared to the Al foam. After the

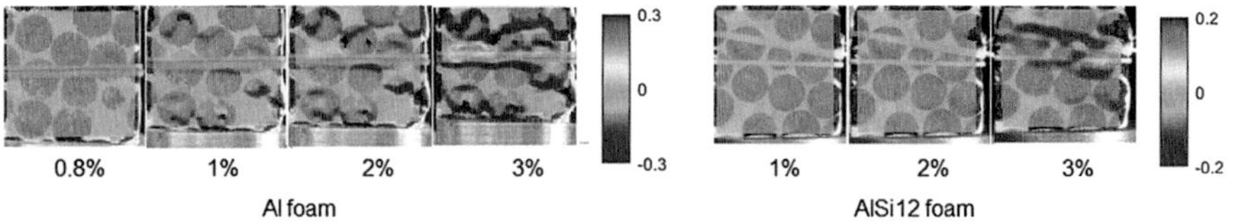

Fig. 5 Surface ε_{yy} strain maps at different macroscopic strains for LECA foams with different matrixes: Al (left) and AlSi12 (right)

appearance of the shear band, the rate of growth in the number of elements in Cluster 2 and 4 increases, while for Cluster 3 decreases. This suggests that in the AlSi12 LECA foam after strain localization the fracture of the matrix becomes important (Fig. 4c).

It is worth noting that according to Fig. 2, the fracture of the cell walls of the LECA particles is the dominant deformation mechanism almost for the entire deformation test. This means that weaker AE signals from other sources are overlapped. This is most likely the reason why the number of elements in the cluster related to plastic deformation in the Al-LECA foam is lower than for the Al foam with ceramic hollow spheres [15].

Conclusion

The effect of the matrix material on deformation and failure mechanisms was investigated in two types of syntactic foams consisting of (i) unalloyed Al matrix or (ii) alloyed Al: Al-12 wt% Si matrix, both reinforced by lightweight expanded clay particles. The differences in deformation mechanism up to ~6% strain were revealed by the help of the ASK analysis of AE data supplemented by imaging techniques. We found that regardless of the matrix material, the identified deformation mechanisms were rather similar, and there was no strain localization in the quasi-linear stage. This resulted in reproducible peak strength with a variation of less than 10%, even for samples with small sample size to LECA particle size ratio. In the case of Al foam, the strain is localized in large volume in multiple deformation bands, which results in reproducible deformation curves even for small samples. After strain localization, the fracture of the matrix material became significant for the AlSi12 foam. This explains the difference in the shape of the stress–strain curves for the AlSi12 foam and the early deterioration of the sample.

Acknowledgements The authors are grateful for the financial support of the Czech Science Foundation under the contract 15-10821S. This paper was supported by the János Bolyai Research Scholarship of the Hungarian Academy of Sciences (Cs. Kádár). M. Knapek gratefully acknowledges partial financial support from the OP RDE, MEYS, grant No. CZ.02.1.01/0.0/0.0/16_013/0001794.

References

1. Taherishargh M, Belova IV, Murch GE, Fiedler T (2014) Low-density expanded perlite–aluminium syntactic foam. Mat Sci Eng A 604:127–134. https://doi.org/10.1016/j.msea.2014.03.003
2. Duarte I, Ferreira JMF (2016) Composite and Nanocomposite Metal Foams. Materials 9:79. https://doi.org/10.3390/ma9020079
3. Gupta N, Rohatgi PK (eds) (2015) Metal matrix syntactic foams: processing, microstructure, properties and applications. DEStech Publications, Lancaster

4. Rohatgi PK, Gupta N, Schultz BF, Luong DD (2011) The synthesis, compressive properties, and applications of metal matrix syntactic foams. JOM 63:36–42. https://doi.org/10.1007/s11837-011-0026-1
5. Balch DK, O'Dwyer JG, Davis GR, Cady CM, Gray GT, Dunand DC (2005) Plasticity and damage in aluminum syntactic foams deformed under dynamic and quasi-static conditions. Mat Sci Eng A 391:408–417. https://doi.org/10.1016/j.msea.2004.09.012
6. Santa Maria JA, Schultz BF, Ferguson JB, Rohatgi PK (2013) Al–Al2O3 syntactic foams – Part I: Effect of matrix strength and hollow sphere size on the quasi-static properties of Al-A206/Al2O3 syntactic foams. Mat Sci Eng A 582:415–422. https://doi.org/10.1016/j.msea.2013.05.081
7. Orbulov IN, Ginsztler J (2012) Compressive characteristics of metal matrix syntactic foams. Compos Part A Appl S 43:553–561. https://doi.org/10.1016/j.compositesa.2012.01.008
8. Szlancsik A, Katona B, Bobor K, Májlinger K, Orbulov IN (2015) Compressive behaviour of aluminium matrix syntactic foams reinforced by iron hollow spheres. Mater Design 83:230–237. https://doi.org/10.1016/j.matdes.2015.06.011
9. Al-Sahlani K, Broxtermann S, Lell D, Fiedler T (2018) Effects of particle size on the microstructure and mechanical properties of expanded glass-metal syntactic foams. Mat Sci Eng A 728:80–87. https://doi.org/10.1016/j.msea.2018.04.103
10. Taherishargh M, Belova IV, Murch GE, Fiedler T (2015) Pumice/aluminium syntactic foam. Mat Sci Eng A 635:102–108. https://doi.org/10.1016/j.msea.2015.03.061
11. Bazzaz Bonabi S, Kahani Khabushan J, Kahani R, Honarbakhsh RA (2014) Fabrication of metallic composite foam using ceramic porous spheres "Light Expanded Clay Aggregate" via casting process. Mater Design 64:310–315. https://doi.org/10.1016/j.matdes.2014.07.061
12. Puga H, Carneiro VH, Jesus C, Pereira J, Lopes V (2018) Influence of particle diameter in mechanical performance of Al expanded clay syntactic foams. Compos Struct 184:698–703. https://doi.org/10.1016/j.compstruct.2017.10.040
13. Blaber J, Adair B, Antoniou A (2015) Ncorr: Open-source 2D digital image correlation Matlab software. Exp Mech 55:1105–1122. https://doi.org/10.1007/s11340-015-0009-1
14. Pomponi E, Vinogradov A (2013) A real-time approach to acoustic emission clustering. Mech Syst Signal Process 40:791–804. https://doi.org/10.1016/j.ymssp.2013.03.017
15. Kádár C, Máthis K, Knapek M, Chmelík F (2017) The effect of matrix composition on the deformation and failure mechanisms in metal matrix syntactic foams during compression. Materials 10:196. https://doi.org/10.3390/ma10020196
16. Yang M, Pan Y, Yuan F, Zhu Y, Wu X (2016) Back stress strengthening and strain hardening in gradient structure. Mater Res Lett 4:145–151. https://doi.org/10.1080/21663831.2016.1153004

Compression Behavior of Low-Pressure Cast AMC Syntactic Foams with High Porosity

Pierre Kubelka, Alexander Martin Matz and Norbert Jost

Abstract This work presents a method for the manufacturing of closed-cell aluminum matrix composite syntactic foams (AMCSFs) using a modified and simplified low-pressure infiltration setup. The influence of different wrought and cast alloys on the compressive behavior of these foams was investigated. Through the use of a variety of different cast and wrought alloys, it was possible to determine the Al matrix' influence on the compressive behavior. The investigated Al*X*-Al_2O_3 syntactic foams were manufactured using hollow alumina spheres with AA1050, AA2024, AA5019, AA7075, and A356 Al alloys in the as-cast state. The results of the manufacturing process and the selected process parameters show a good dispersion of the spheres within the Al*X* matrix with a typical near randomly close-packed structure at the same time. The high-strength Al alloys AA2024 and AA7075 lead to a very brittle deformation behavior of the foams in the compression tests, with a strongly oscillating behavior plateau and relatively low plateau stress level. In contrast, the low- and mid-strength alloys AA1050, AA5019, and A356 show a more ductile behavior with less oscillation at a higher plateau stress level. By the five different combinations of the material partners, it was possible to make a statement about the ductility–strength relation of metallic syntactic foams in dependence on the base matrix.

Keywords Syntactic foams · Compressive behavior · Alumina spheres · Aluminum alloys · Casting

Introduction

Aluminum foams have been undergoing a fast-paced development in the last decades regarding their mechanical properties related to lightweight design. Due to the combination of aluminum as the base material with a porous structure, it is possible to provide a structure with multifunctional properties such as a high specific strength,

---

P. Kubelka (✉) · A. M. Matz · N. Jost
Institute for Materials and Material Technologies (IMMT), Pforzheim University, Tiefenbronner Straße 65, 75175 Pforzheim, Germany
e-mail: pierre.kubelka@hs-pforzheim.de

N. Dukhan (ed.), *Proceedings of the 11th International Conference on Porous Metals and Metallic Foams (MetFoam 2019)*, The Minerals, Metals & Materials Series, https://doi.org/10.1007/978-3-030-42798-6_11

good energy-absorbing capabilities, and a good damping behavior [1–3]. In this field, aluminum matrix composite syntactic foams (AMCSFs) have been regaining a lot of interest recently, owing to the possibility of cheaper production and materials. These closed-cell foam structures combine the two different aspects of minimizing the material's density as a foam with a simultaneous reinforcement of the Al-base material by the use of porous and hollow discontinuous reinforcements, respectively. However, this makes them a more suitable lightweight material in comparison to non-reinforced open- and closed-cell aluminum foams.

A wide range of different material combinations and structural effects of AMCSFs have been studied regarding their manufacturing and compressive properties. The investigated Al-base materials are ranging from pure Al [4–10], AlCu alloys [7, 11], AlMg alloys [7–9, 12], AlSi alloys [5–7, 13–19] up to AlZn alloys [7, 10]. The spherical reinforcements are typically represented by ceramics like alumina [6, 7, 9, 11], silicon carbide [18], perlite [17], glasses [8, 13], and mixtures of silica and alumina with other additives [4, 5, 7, 10, 12, 14–16, 19, 20] where the geometric parameters of the spheres can range from several microns up to several millimeters. Since these previous investigations were only covering individual Al-base and reinforcement matrix combinations, a comprehensive statement about the impact of the Al-base matrix on the compressive behavior of the AMCSFs cannot be given.

This study follows the approach of generating a holistic understanding of the impact of the Al-base matrix on the compressive behavior of AMCSFs with hollow alumina spheres. For this, mm-sized hollow alumina spheres combined with different commercial Al wrought alloys (AA1050, AA2024, AA5019, AA7075) and one cast alloy (A356) are being investigated. Together with a simplified and modified low-pressure infiltration casting method, it is possible to provide AMCSFs with different Al-base matrices manufactured under simplified constant conditions with unchanged geometric parameters. Undergoing compressive testing, it is possible to provide information on the impact of the base material onto the general deformation behavior regarding the strength and ductility of the AMCSFs.

Experimental Methods

Raw Materials

The materials used to produce the AMCSFs are represented by hollow alumina spheres and different Al wrought and cast alloys. The hollow Al_2O_3 spheres (hollomet GmbH, Dresden, GER) were made of commercial C799 alumina and had a mean outer diameter of 3.83 mm with a shell thickness of 150 μm. The only investigated cast alloy is the commercially most used alloy A356 (AlSi7Mg0.3) with the ability to be artificially aged. The investigated Al wrought alloys can be divided into natural hard alloys and alloys with the possibility to be artificially and naturally aged. In terms of natural hard alloys, AA1050 (Al99.5) and AA5019 (AlMg5) were

Table 1 Chemical composition of the different Al wrought and cast alloys used for the investigations (in wt%)

Alloy	Si	Fe	Cu	Mg	Mn	Zn	Ti	Cr	Al	Rest
AA1050	0.214	0.353	0.006	0	0.004	0	0.015	0.011	99.344	0.052
AA2024	0.394	0.235	4.369	1.649	0.798	0.123	0.007	0.007	92.300	0.076
AA5019	0.418	0.298	0.029	5.156	0.210	0.018	0.096	0.096	93.688	0.075
AA7075	0.322	0.262	1.692	2.480	0.096	5.666	0.194	0.194	89.200	0.061
A356	7.330	0.110	0.003	0.360	0.001	0	0.110	0.001	91.938	0.146

investigated. For alloys with the ability of natural aging, AA2024 (AlCu4Mg1) and AA7075 (AlZn5.5MgCu) were studied. The composition of the investigated alloys was analyzed using spark emission spectroscopy and is shown in Table 1. Elementary components indicated with a value of zero did not reach the lower threshold value of the measuring range.

Manufacturing

The AMCSFs were manufactured using a modified and simplified low-pressure casting method. The complete setup of the process with the prepared mold is shown in Fig. 1. The alumina spheres were placed into a boron-nitride-coated stainless steel mold with a square cross section of 20 × 20 mm^2 and were gently tapped in order to achieve a randomly close-packed structure [21]. To maintain the dense packaging and to prevent floating of the spheres while being infiltrated by the molten aluminum, an AISI304 grid with a mesh size of 2 mm was clamped right above the package. The mold with the package was then heated up to a temperature of 600 °C and given into the casting machine. The infiltration of the alumina sphere package was done using an Indutherm VC500 vacuum-casting machine. The modified process used can be divided into the three different steps of *melting*, *casting/infiltration,* and *cooling*.

1. The *melting* of the Al alloys was performed inductively in Ar atmosphere under atmospheric normal pressure at a temperature of 1,000 °C. After finishing the process of melting, the temperature was held for several further seconds to gain uniform distribution of the elements and temperature within the melt.
2. The *casting* and *infiltration* were performed as a sequence of these two steps. The *casting* was done by pouring the melt onto the package without using any further pressurization of the Ar atmosphere. This led to a complete coverage of the cross section of the sphere package, enabling an uniform infiltration. The *infiltration* itself was set in after a short delay of 0.3 s through the pressurization of the melt with 1.5 bar by the Ar atmosphere. The pressure was being held up for a minimum of 5 s to maintain the complete infiltration of the package until the solidification of the Al*X* matrix took place.

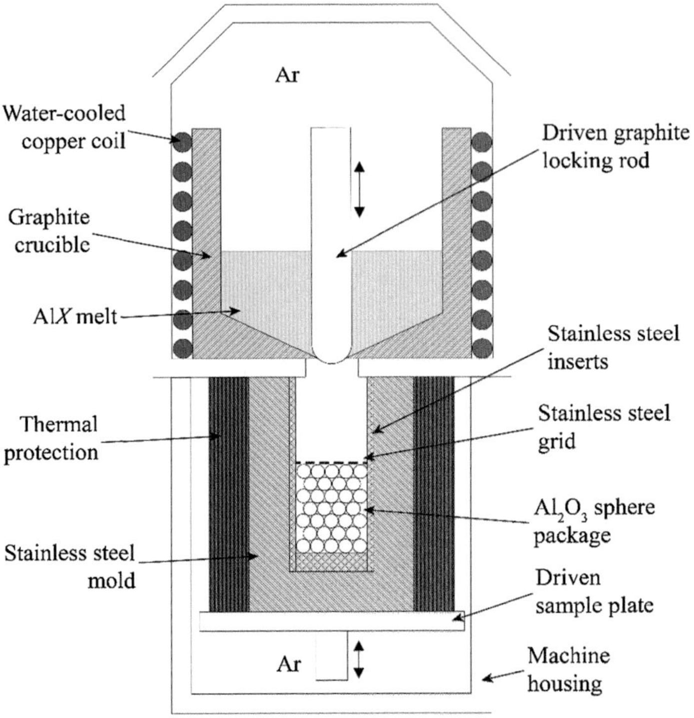

Fig. 1 Schematic setup of the modified and simplified low-pressure casting method within an Indutherm VC500 casting machine. The figure shows the process in the *melting* stage right before the *casting* and *infiltration*

3. The sample was cooled down within the machine and the mold to approximately 300 °C within 5 min. Afterward, the sample was demolded and cooled to RT in air.

The as-gained Al*X*-Al_2O_3 syntactic foam samples were finally machined to achieve defined geometric parameters. The foams were cut with a metal-bonded diamond cutting wheel in an ATM Brilliant 220 wet abrasive cutting machine to obtain samples with a length of 20 mm, which led to overall outer dimensions of 20 × 20 × 20 mm^3. The separated surfaces of the samples were finally wet grinded with SiC abrasive paper to remove the remaining burrs and to smoothen the surface.

Compressive Testing

Prior to the compressive testing the relative densities ρ_{rel} of the samples for comparison purposes were calculated. For this, the outer dimensions and the weight of the samples were measured in order to calculate the density of the foams ρ_{foam}. This

Fig. 2 An example of the macrostructure of the as-cast AMCSFs

density was set in relation to the Al alloy's density ρ, resulting in the relative density of the foam (Eq. 1). With this procedure only the densities of the Al alloys were considered and the sphere's density was assumed to be identical. This small error could be neglected since identically produced samples were being compared in this investigation.

$$\rho_{\text{rel}} = \frac{\rho_{\text{foam}}}{\rho} \tag{1}$$

The compressive tests were performed using a Galdabini QUASAR 250 universal test rig. The compressive forces were applied onto the as-cast samples by using two hardened precision-grinded steel plates. The resulting deformation direction was perpendicular to the cut and ground sides and parallel to the casting direction. An example of the machined macrostructure of the as-cast AMCSFs is given in Fig. 2. At the beginning of the procedure, the samples were pre-compressed with a force F of 10 N. The testing was done using a displacement rate of 4 mm/min which corresponds to a strain rate of $3.33 \times 10^{-3}\ \text{s}^{-1}$. The force–displacement data was recorded with a sampling rate of $100\ \text{s}^{-1}$ to obtain the effective stress–strain curves for each individual sample until the maximum strain of 0.7. A force of 150 kN was set as the termination criterion for the tests.

Metallographic Analysis

To support the results of the deformation tests, the microstructures of AMCSFs were investigated. Therefore, slices perpendicular to the deformation direction were machined and embedded in ClaroCit cold-mounting resin. The embedded specimens were ground and polished with diamond particles up to 1 μm. *Barker* etchant was used to develop the microstructure of the as-cast Al alloys. The optical characterization was performed using a Leica DMI5000M inverted light microscope.

Results and Discussion

Manufacturing and Microstructure

The low-pressure infiltration of the sphere package shows a relatively good sphere distribution with a nominal relative density of $\rho_{rel} \approx 56.5\%$, independent of the alloy investigated. The randomly close-packed structure, as a direct result of the tapping of the mold while placing the spheres and preventing the floating with a clamped stainless steel grid, remains almost unchanged in the as-cast samples. The microstructures of the different alloys have developed almost identically with a fine-grained and homogeneous microstructure within the samples (Fig. 3a, b). Only the AA5019 and AA7075 alloys are showing a gradation in the grain size from the inside to the outside. This is due to the lower solidification temperatures compared to the

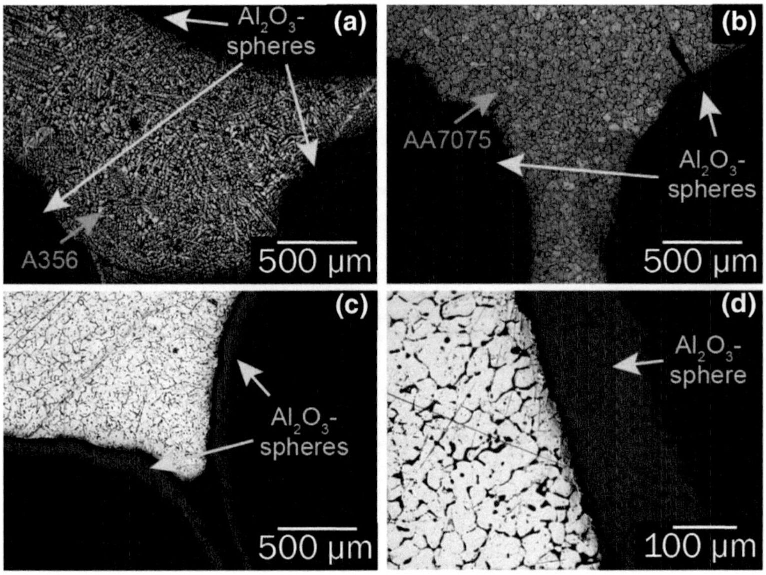

Fig. 3 Microstructures of different alloys as well as the filling and interfacial behavior between the constituents **a** microstructure of an as-cast A356 alloy, **b** microstructure of an as-cast AA7075 alloy, **c** infiltration behavior of the different Al alloys in spherical contact areas with voids, **d** interlocking between the solidified base matrix and the spherical reinforcement with small cavities

AA1050, AA2024, and A356 alloys. Looking at the micrographs of the ground and polished samples, an overall good interlocking between the Al*X* matrices and the alumina spheres can be detected (Fig. 3c, d). Small numbers of defects within the foam structures can also be noticed. In regions of interconnecting spheres, cavities with prematurely solidified melt fronts can be identified, which has also been reported by other researchers [17, 20]. Combined with small numbers of randomly distributed cavities at the interfaces, it follows that the used process parameters and pre-treatments of the materials do not yet correspond to the optimum for the production of cavity-free AMCSFs with this method. This problem can be solved by positively affecting the wetting behavior between Al*X* melt and the spheres in the infiltration stage. Due to the poor wetting of melted aluminum and its alloys with alumina, several changes can be made to minimize the contact angle between the two constituents. The change of the Al alloy composition under consideration of the Mg content as well as changes in the infiltration or mold temperatures can lead to better results [22, 23].

Compressive Behavior

Typical stress–strain curves for the different Al alloys investigated are shown in Fig. 4a. Compared to other investigations regarding melt infiltrated AMCSFs with mm-sized ceramic spheres [9, 17, 18, 20, 24, 25] an overall more brittle and oscillating behavior can be observed. This is due to the combination of the fast solidification of the melts while infiltrating the sphere package, the present microstructural conditions as well as the natural brittle behavior of the alumina spheres. Since only the influence of the Al alloys on the compressive behavior was investigated, the influence of the alumina spheres is not discussed below.

The melt solidifies within seconds and cools down to RT in less than 10 min, resulting in the fine-grained structure in the as-cast state. Residual stresses in the base matrix around the spheres occur during the rapid cooling because of the mismatch of the coefficients of thermal expansion of the Al*X* matrix to the alumina [26]. Comparing the five base materials with each other, a clear difference between the high-strength alloys (AA2024 and AA7075) and the low- to mid-strength alloys (AA1050, AA5019, and A356) can be observed. The high-strength alloys AA2024 and AA7075 can both be naturally and artificially aged [27, 28] resulting in a brittle matrix. Since the as-cast state provides relatively undefined microstructures, it must be assumed that partially artificial and natural aged conditions are present in the materials. The higher plastic collapse stresses σ_{pc} and the low plateau stresses σ_{pl} with a strongly oscillating plateau support this statement. The low- and mid-strength alloys AA1050, AA5019, and A356 are providing lower plastic collapse stresses with partially increased plateau stress levels (AA5019 and A356). These alloys have a much more ductile nature as the high-strength alloys as they are not able to be naturally aged. There are no finely dispersed sub-micron precipitations which inhibit a ductile behavior. Considering the maximization of the ductility with a simultaneous

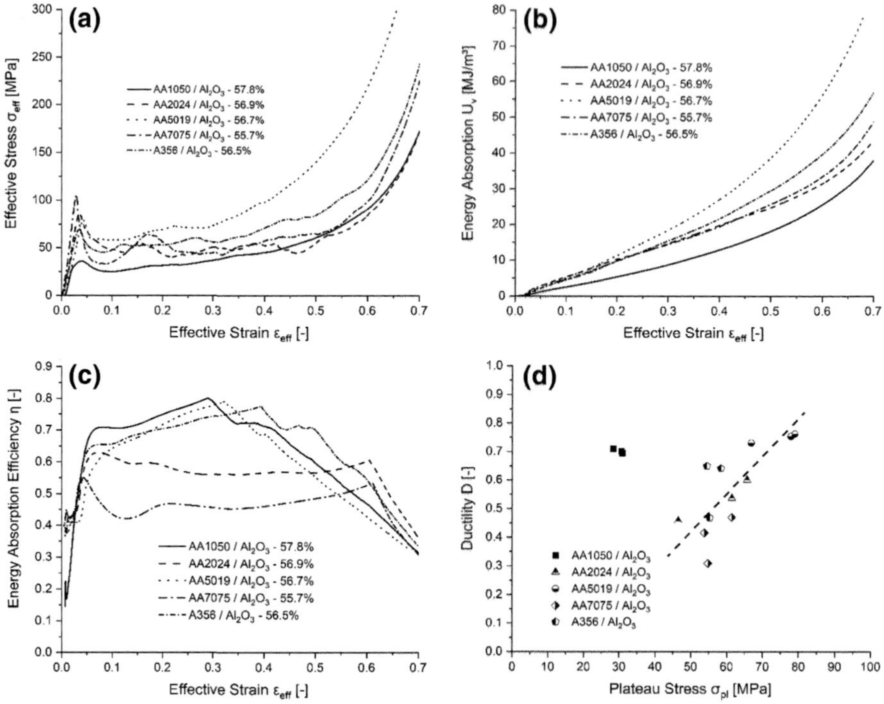

Fig. 4 **a** Effective stress–strain curves σ_{eff} (ε_{eff}) of all different AMCSFs investigated, **b** volumetric energy absorption curves $U_V(\varepsilon_{eff})$ of all different AMCSFs investigated, **c** energy absorption efficiency curves $\eta(\varepsilon_{eff})$ of all different AMCSFs investigated, **d** calculated ductility *D* versus the plateau stress σ_{pl} of all different AMCSFs investigated

high strength, the AA5019 alloy shows a pretty good behavior. This natural hard alloy cannot be artificially or naturally aged which is why a good ductility with a moderately high basic strength at the same time is maintained.

These different behaviors can also be verified by the energy-absorbing behavior. The energy absorption efficiency η shall be used for this assessment. It is defined as the actual absorbed energy under the stress–strain curve divided by the energy absorbed by an ideal absorber with the maximum stress σ_{max} reached up to the strain ε [29]. The energy absorption U_v [30] was calculated by Eq. 2 and the energy absorption efficiency was calculated by Eq. 3. The typical results of the energy absorption and efficiency for the different AMCSFs are displayed in Fig. 4b, c.

$$U_v = \int \sigma(\varepsilon)d\varepsilon \tag{2}$$

$$\eta = \frac{\int \sigma(\varepsilon)d\varepsilon}{\sigma_{max} \cdot \varepsilon} \tag{3}$$

When considering the high-strength alloys AA2024 and AA7075, a pretty low overall efficiency can be observed. The massive stress drop from the plastic collapse stress down to a pretty low plateau stress level leads to a poor energy-absorbing performance. The buildup of narrow deformation bands with repetitive failure and rebuilding at higher strains leads to numerous small material breakouts within the foam until the densification of the foams. A very weak load distribution over all spheres onto the whole cross-sectional area of the foam is the result. Evaluating the energy absorption efficiencies of the three low- and mid-strength alloys, a higher overall efficiency can be observed. Their stress drops down to the plateau stress levels are smaller, positively affecting the energy absorption efficiency. These alloys are more ductile allowing a better load distribution within the matrix onto a higher number of ceramic spheres. Locally higher strains can be better cushioned which results in a much smoother plateau phase with less oscillation and break-ins. To make the influence of the Al*X* matrix on the ductility of the investigated AMCSFs tangible, a method for calculating such has been developed. This method describes the ductility D of these structures as the quotient of the minimum stress σ_{min} in the plateau phase (II) and the plastic collapse stress σ_{pc} (see Eq. 4). The plastic collapse stress is defined as the peak stress at which AMCSF transitions from the pseudo-elastic phase (I) to the plateau phase (II). Figure 5 displays the used parameters within the different phases summarized in a schematic plot.

$$D = \frac{\sigma_{min}}{\sigma_{\text{pc}}} \tag{4}$$

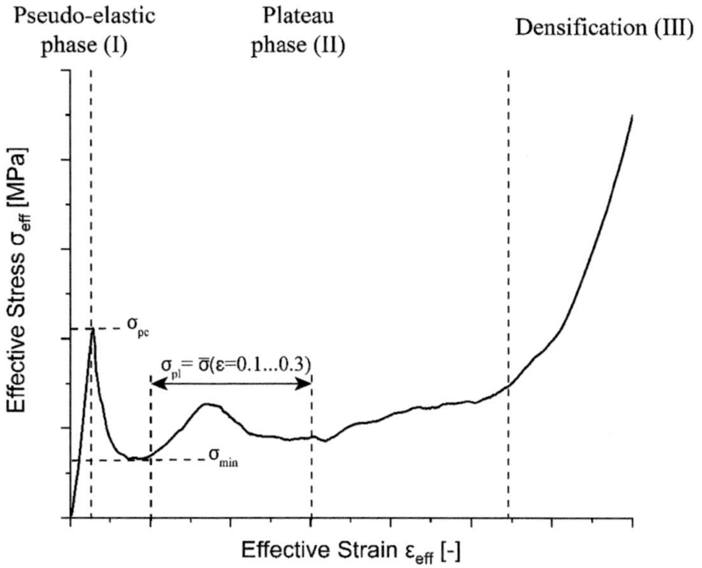

Fig. 5 Schematic plot of a brittle AMCSF showing the different compressive phases and the parameters used for the assessment of the overall ductility–strength behavior

Plotting the calculated ductility *D* above the plateau stress (mean stress between the strain of 0.1 and 0.3) of each individual investigated AMCSF sample results in a cloud of points which shows a linear relation between the ductility and strength of the foams as a function of the Al alloy (see Fig. 4d). Thereby, it can be shown that the selected base material and its condition have a substantial influence on the compressive deformation behavior and the strength of these structures. Multiphase high-strength Al*X* matrices (AA2024, AA7075) are negatively affecting the deformation behavior, while low- (AA1050) to mid-strength (AA5019, A356) alloys, with single- or dual-phase matrices, are positively affecting the ductility. The single-phase AA1050 (nearly pure Al) ensures that even softening with a simultaneous high ductility of the deformation behavior can occur. Multi-component alloys are therefore indispensable to achieve an increased overall strength of AMCSFs.

Conclusion

A modified and simplified low-pressure casting method was used to manufacture AMCSF samples with different Al*X* matrices to determine the impact of the base material on their compressive behavior. The produced samples were showing a near randomly dense-packed structure of the alumina spheres with small defects in the material's matrix (cavities and voids). The compressive tests have shown that the base materials have a significant influence on the deformation behavior of the AMCSF itself. High-strength alloys (AA2024, AA7075) have, in view of the as-cast states with an undefined phase composition, a negative impact onto the ductility and strength of the foams. Localized high strains ensure that an oscillating deformation behavior with a low plateau stress is achieved. Low- (AA1050) to mid-strength alloys (AA5019, A356) are more ductile, regarding the deformation behavior of the foam, because they are not as strongly affected by aging effects as the high-strength alloys. In comparison, it is possible to maximize the ductility and strength with these materials in the as-cast state. The developed method to calculate the ductility and plotting it above the plateau stress has demonstrated that not only the base material itself has an impact on the deformation behavior, but also the microstructural condition in which it is (as-cast, aged, …) and the elemental composition of the alloys.

Acknowledgements The authors gratefully acknowledge financial support by the Europäischer Fonds für regionale Entwicklung (EFRE) and the Ministerium für Wissenschaft, Forschung und Kunst Baden-Württemberg within the research center ZAFH InSeL and the financial support through the program "Mittelbau" at HAW by the Ministerium für Wissenschaft, Forschung und Kunst Baden-Württemberg.

References

1. Ashby MF et al (2000) Metal foams: a design guide. Butterworth Heinemann, Burlington
2. Gibson LJ, Ashby MF (1997) Cellular solids: structure and properties. Cambridge University Press, Cambridge
3. Banhart J (2013) Light-metal foams—history of innovation and technological challenges. Adv Eng Mater 15(3):82–111
4. Xia X et al (2014) Compressive properties of closed-cell aluminum foams with different contents of ceramic microspheres. Mater Des 56:353–358
5. Szlancsik A et al (2017) On the effective Young's modulus of metal matrix syntactic foams. Mater Sci Tech 33(18):2283–2289
6. Su M et al (2019) Compressive properties of aluminum matrix syntactic foams prepared by stir casting method. Adv Eng Mater 21:1900183
7. Orbulov IN, Májlinger K (2013) Description of the compressive response of metal matrix syntactic foams. Mater Des 49:1–3
8. Lin Y et al (2017) Microstructure and strength correlation of pure Al and Al-Mg syntactic foam composites subject to uniaxial compression. Mat Sci Eng A 696:236–247
9. Castro G et al (2013) Compression and low-velocity impact behavior of aluminum syntactic foam. Mat Sci Eng A 578:222–229
10. Balch DK et al (2005) Plasticity and damage in aluminum syntactic foams deformed under dynamic and quasi-static conditions. Mat Sci Eng A 391:408–417
11. Santa Maria JA et al (2013) Al-Al_2O_3 syntactic foams—part I: effect of matrix strength and hollow sphere size on the quasi-static properties of Al-A206/Al_2O_3 syntactic foams. Mat Sci Eng A 582:415–422
12. Palmer RA et al (2007) Pressure infiltrated syntactic foams—process development and mechanical properties. Mat Sci Eng A 464:85–92
13. Wright A, Kennedy A (2017) The processing and properties of syntactic Al foams containing low cost expanded glass particles. Adv Eng Mater 19(11):1600467
14. Orbulov IN (2013) Metal matrix syntactic foams produced by pressure infiltration—the effect of infiltration parameters. Mat Sci Eng A 583:11–19
15. Tao XF et al (2009) Al matrix syntactic foam fabricated with bimodal ceramic microspheres. Mater Des 30:2732–2736
16. Tao XF, Zhao YY (2012) Compressive failure of Al alloy matrix syntactic foams manufactured by melt infiltration. Mat Sci Eng A 549:228–232
17. Taherishargh M et al (2015) On the particle size effect in expanded perlite aluminium syntactic foam. Mater Des 66:294–303
18. Luong DD et al (2013) Development of high performance lightweight aluminum alloy/SiC hollow sphere syntactic foams and compressive characterization at quasi-static and high strain rates. J Alloy Compd 550:412–422
19. Birla S et al (2017) Effect of cenosphere content on the compressive deformation behaviour of aluminum-cenosphere hybrid foam. Mat Sci Eng A 685:213–226
20. Myers K et al (2015) Quasi-static and high strain rate response of aluminum matrix syntactic foams under compression. Compos Part A 79:82–91
21. Jaeger HM, Nagel SR (1992) Physics of the granular state. Science 255(5051):1523–1531
22. Ip SW et al (1993) Wetting behaviour of aluminium and aluminium alloys on Al_2O_3 and CaO. J Mater Sci Lett 12(21):1699–1702
23. Klinter AJ et al (2008) Wetting of pure aluminum and selected alloys on polycrystalline alumina and sapphire. Mat Sci Eng A 495:147–152
24. Orbulov IN et al (2019) Compressive characteristics of bimodal aluminium matrix syntactic foams. Compos Part A 124:105479
25. Katona B et al (2019) Compressive characteristics and low frequency damping of aluminium matrix syntactic foams. Mat Sci Eng A 739:140–148
26. Arsenault RJ, Shi N (1986) Dislocation generation due to differences between the coefficients of thermal expansion. Mat Sci Eng 81:175–187

27. Guinier P (1938) Structure of age-hardened Aluminium-Copper alloys. Nature 142(3595):569–570
28. Berg LK et al (2001) GP-zones in Al-Zn-Mg alloys and their role in artificial aging. Acta Mater 49(17):3443–3451
29. Baumeister J et al (1997) Aluminium foams for transport industry. Mater Des 18(4–6):217–220
30. Paul A, Ramamurty U (2000) Strain rate sensitivity of a closed-cell aluminum foam. Mat Sci Eng A 281(1–2):1–7

Preparation and Quasi-static Compression Behavior of Steel Matrix Syntactic Foams

Quanzhan Yang, Bo Yu, Yanpeng Wei, Guang Hu, Zhiquan Miao and Jingchang Cheng

Abstract Metal matrix syntactic foams have been attracting considerable attention in recent years due to their potential applications in lightweight structures and energy absorption. In this paper, CF-8 cast austenitic stainless steel with hollow Al_2O_3 spheres' syntactic foams was prepared by infiltration casting technology. The effects of preheat temperature of ceramic mold and diameter of hollow ceramic spheres on the forming ability of infiltration casting of syntactic foams were investigated and the results showed that suitable preheat temperature of ceramic mold is a necessary prerequisite to produce the syntactic foams. And then, the quasi-static compressive behavior of these foams was studied by experimental and numerical simulation and the energy absorption capacity is evaluated.

Keywords Steel matrix syntactic foams · Infiltration casting · Quasi-static compression behavior · Numerical simulation · Energy absorption

Introduction

Metal matrix syntactic foams consisting of a continuous metal matrix embedding hollow spheres have great potential applications in lightweight structures and energy absorption [1–3]. Currently, many kinds of metal matrix syntactic foams have been developed, such as Al matrix syntactic foams [4–8], Mg matrix syntactic foams [9], Zn matrix syntactic foams [10], and so on. Among them, steel matrix could provide higher mechanical properties compared to other metal matrix. However, investigations about steel matrix syntactic foams are very few due to the challenge that lies in the characteristics of steel, especially the high melting point of steel which rules out or complicates some production processes employed routinely in syntactic metal

Q. Yang · B. Yu (✉) · Y. Wei · Z. Miao · J. Cheng
Shenyang Research Institute of Foundry Co. Ltd., State Key Laboratory of Light Alloy Casting Technology for High-End Equipment, Shenyang 110022, China
e-mail: yub@chinasrif.com

G. Hu
Xi'an Jiao Tong University, Xi'an 710049, China

N. Dukhan (ed.), *Proceedings of the 11th International Conference on Porous Metals and Metallic Foams (MetFoam 2019)*, The Minerals, Metals & Materials Series, https://doi.org/10.1007/978-3-030-42798-6_12

Table 1 Chemical composition of CF-8 cast austenitic stainless steel in wt%

C	Si	Mn	Cr	Ni	Fe
0.02	1.0	1.6	19.1	10.2	Balance

foam. In this paper, an infiltration casting technology was developed for preparing steel matrix syntactic foams, in which ASTM CF-8 cast austenitic stainless steel and hollow Al_2O_3 spheres were chosen as matrix and preformed particles, respectively. The influences of ceramic mold preheating temperature and hollow ceramic sphere's diameter on infiltration length were investigated. In addition, the steel matrix syntactic foam's compression stress–strain behavior was studied by experimental and numerical simulation.

Materials and Methods

Infiltration Casting Experimental Procedures

ASTM CF-8 cast austenitic stainless steel was used as the matrix metal to create syntactic foams and the chemical composition was listed in Table 1. Three types of high-purity Al_2O_3 hollow spheres were used as the pore-forming material, the average outer diameters size was about Ø3.11 mm, Ø3.97 mm, and Ø4.79 mm.

During the process of preparation, hollow Al_2O_3 spheres were embedded in an alumina ceramic mold to form a random loose packing structure. Meanwhile the stainless steel was heated above its melting temperature and then the molten steel was poured into the ceramic mold that has been preheated. The molten steel would fulfill the spaces between the hollow spheres. After the process of solidification, clearing getting system, etc., steel matrix syntactic foams were obtained. Then, the samples were heat-treated at 1050 °C temperature for 2 h and water quenched as followed (Fig. 1).

Experimental and Numerical Simulation of Quasi-static Compression Behavior

The average diameter of the spheres in the syntactic foams selected for quasi-static compression test was about 4.79 mm and the dimensions for compression test samples were about 20 mm × 20 mm × 30 mm. A 1 kN universal testing machine was used to evaluate syntactic foams' compression behaviors. The loading strain rate was fixed at 1.25 mm/min. The finite element analysis software of ABAQUS was applied to calculate the compression response, a full 3D model was constructed via the secondary development of ABAQUS by using Python language, as shown in Fig. 2.

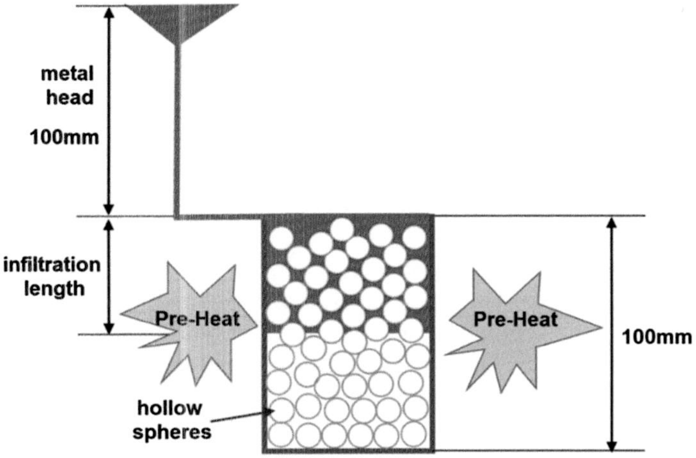

Fig. 1 Schematic illustration of infiltration casting process

Fig. 2 3D geometric model for quasi-static compression test

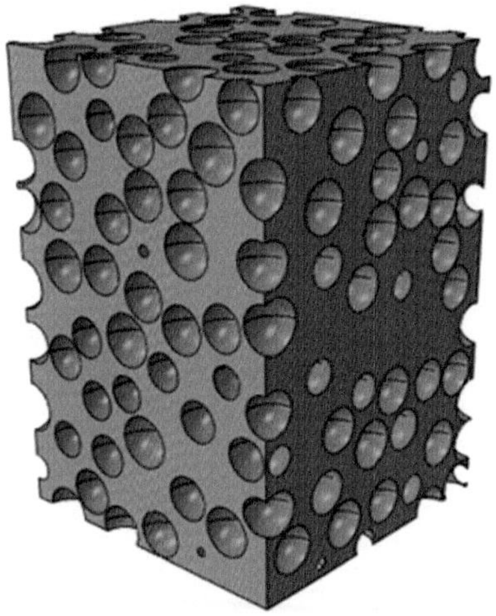

The basic mechanical properties of the CF-8 cast austenitic stainless steel were listed in Table 2, and Young's modulus was 190 GPa and Poisson's ration was 0.29.

Table 2 Plasticity data of the matrix metal

True stress/MPa	190	216	228	236	242	247
True strain	0	0.2	0.4	0.6	0.8	1.0

Results and Discussion

Infiltration Casting Process

It can be seen from Fig. 3 that the infiltration length of the sample increases when the preheat temperature of ceramic mold increases, no matter how the hollow sphere's size was changed. The temperature difference between molten steel and hollow spheres is large and the heat exchange is intense when the ceramic mold is at a lower preheat temperature. The liquid metal around the spheres will soon cool down to form a solidified shell, which in turn will cause the infiltration process to stop and the incomplete infiltration of the syntactic foams will occur.

For different diameters of hollow Al_2O_3 spheres that are about 3.11 mm, 3.97 mm, and 4.79 mm, respectively, the infiltration effect is quite different. The smaller the diameter of spheres, the larger the flow resistance, and the larger the heat exchange area between the molten steel and the spheres. Thus the molten steel tends to easily solidify and block the channel during the filling process, shortening the infiltration length of the synthetic foam. The ceramic mold charged by spheres with a diameter of 4.79 mm in which the foams formed must be preheated to at least 1000 °C, and with the diameter of 3.11 mm, 3.97 mm in which the foams prepared must be preheated to at least 1200 °C to prevent incomplete infiltration. Based on the findings, CF-8 cast austenitic stainless steel with hollow Al_2O_3 spheres' syntactic foams with different outer diameter sizes of spheres was prepared successfully by infiltration casting technology, as shown in Fig. 4.

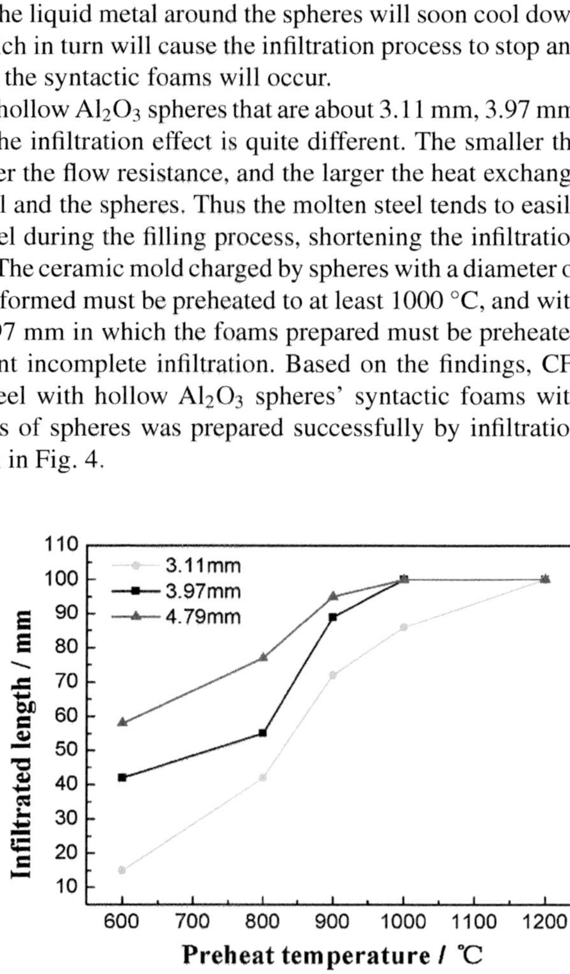

Fig. 3 Infiltration length of molten steel into the spaces between the hollow spheres with different preheat temperatures

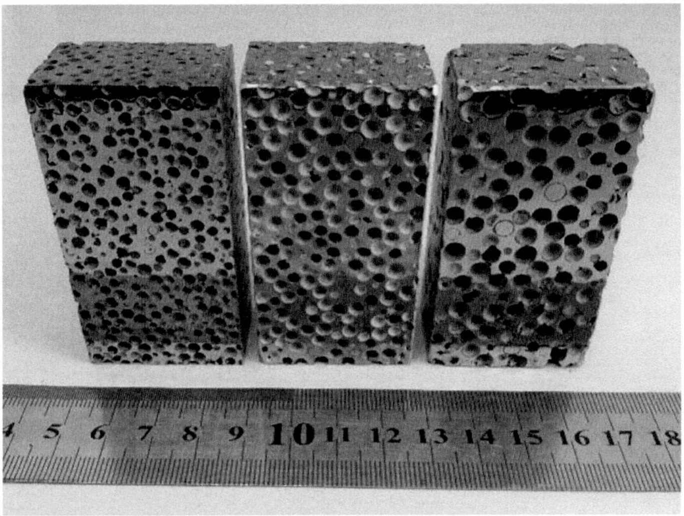

Fig. 4 The samples of steel matrix syntactic foams

Quasi-static Compression Behavior

The experimental and simulated quasi-static compression stress–strain curves of steel matrix syntactic foams have shown typical three stages in Fig. 5, which contain initial linear elastic stage (Oa, Oa′), platform stress stage (ab, a′b′), and densification stage (bc, b′c′). The results indicate that steel matrix syntactic foams possess similar compression behavior to those of conventional metal foams reported in previous studies [11, 12]. In linear elastic stage, the stress is small, the steel matrix syntactic foams present elastic deformation, and the stress–strain relationship is linear. With

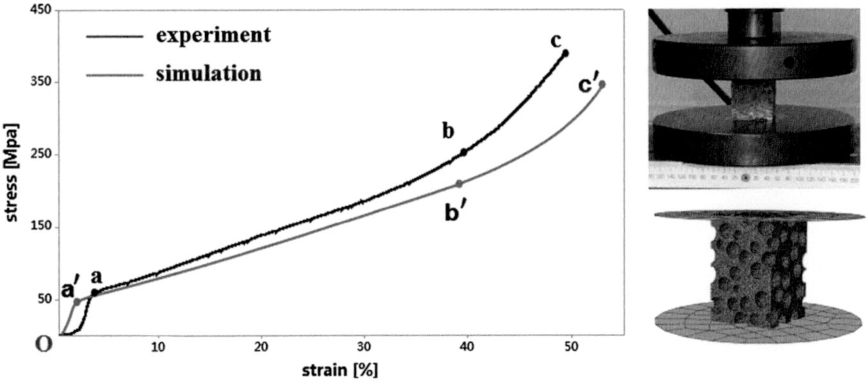

Fig. 5 Compressive stress–strain curves of steel matrix syntactic foams

the increase of load, the syntactic foams enter the stress plateau stage, and the cell wall of the syntactic foams presents plastic deformation, and finally collapses. The stress does not increase or increase slowly at the platform stress stage, and the strain increases rapidly. When the load increases further, the syntactic foams enter the densification stage, the cell wall of the syntactic foams is compacted, the strain increases slowly, and the stress increases rapidly.

The comparison between compression test curves and simulation curves showed that the overall of the two curves has a similar trend and the results showed that simulated quasi-static compression stress–strain curves were in good agreement with the experimental value and the error of compression stress is less than 15%. In the platform stress stage, the test curve stress increases rapidly, which is different from the numerical simulation value, and it was thought that the presence of hollow alumina spheres will increase the stress of synthetic foams, but the enhancement effect of alumina hollow spheres is not considered in the numerical model.

Energy Absorption

The energy absorption of the syntactic foam is obtained by calculating the area under the stress–strain curve up to its densification strain. The energy absorbed per unit volume W_V and energy absorbed per unit mass W_m are defined as follows:

$$W_V = \int_0^{\varepsilon_D} \sigma d\varepsilon \tag{1}$$

$$W_m = \int_0^{\varepsilon_D} \sigma d\varepsilon / \rho \tag{2}$$

where ε_D is the densification strain, σ is the compressive stress, and ε is the compressive strain. For the compressive deformation of foams, densification strain point is an important criterion for calculating energy absorption capacity, that is, the slope $({}^{d_\sigma}/_{d_\varepsilon})$ of compressive stress–strain curve increases obviously after the plateau region, as shown in Fig. 6. The energy absorbed per unit volume is 54.47 MJ/m^3 and energy absorbed per unit mass is 14.964 kJ/kg for syntactic sample with 40.293% densification strain.

Conclusions

In summary, an infiltration casting method was developed to prepare steel matrix syntactic foams, which contain different outer diameters of hollow Al_2O_3 spheres that are about Ø3.11 mm, Ø3.97 mm, and Ø4.79 mm, respectively. The infiltration process and compression behavior were studied and the energy absorption capacity

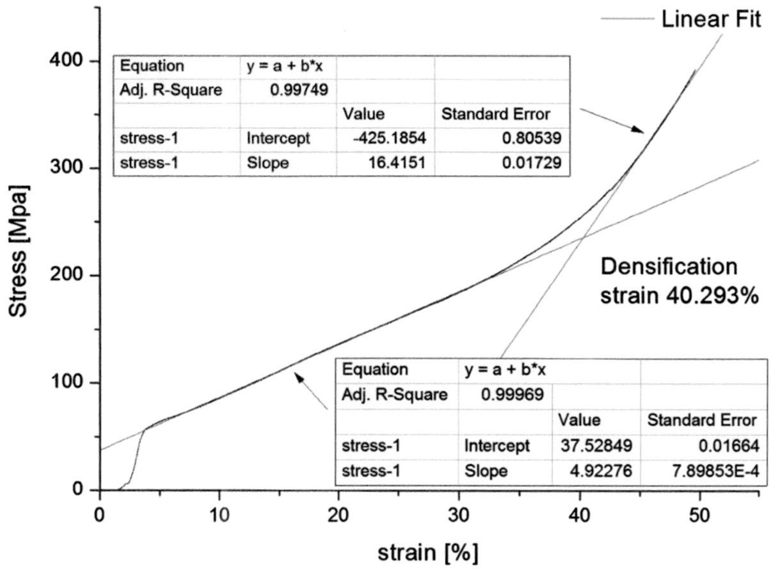

Fig. 6 Densification strain of steel matrix syntactic foams

is evaluated. The results showed that suitable preheat temperature of ceramic mold is a necessary prerequisite to produce syntactic foams. The simulation and experimental stress–strain curve of syntactic foam exhibits the typical "three stages" in the quasi-static compression. The results showed that simulated quasi-static compression stress–strain curves are in good agreement with the experimental value. The energy absorbed per unit volume is 54.47 MJ/m^3 and energy absorbed per unit mass is 14.964 kJ/kg for syntactic sample with 40.293% densification strain.

Acknowledgements This project was supported by Natural Science Foundation of Liaoning Province of China (No.20180550876) and State Key Laboratory of Light Alloy Foundry Technology for High-end Equipment (No. LACT-001). The authors would like to acknowledge these organizations for financial support.

References

1. Rabiei A, Vendra LJ (2009) A comparison of composite metal foam's properties and other comparable metal foams. Mater Lett 63(5):533–536
2. Neville BP, Rabiei A (2008) Composite metal foams processed through powder metallurgy. Mater Design 29(2):388–396
3. Weise J, Baumeister J, Yezerska O, Salk N, Silva GBD (2010) Syntactic iron foams with integrated microglass bubbles produced by means of metal powder injection moulding. Adv Eng Mater 12(7):604–608

4. Ferguson JB, SantaMaria JA, Schultz BF (2013) Al–Al_2O_3 syntactic foams—Part II: Predicting mechanical properties of metal matrix syntactic foams reinforced with ceramic spheres. Mater Sci Eng A 582(2):423–432
5. Kádár C, Máthis K, Orbulov IN (2016) Monitoring the failure mechanisms in metal matrix syntactic foams during compression by acoustic emission. Mater Lett 173:31–34
6. Lin Y, Zhang Q, Liu T, Wang H, Lu J, Ye Y, Zheng K (2019) Sol-gel MgO coating on glass microspheres for inhibiting excessive interfacial reaction in Al-Mg matrix syntactic foam. J Alloys Compd 798(25):59–66
7. Katona B, Szlancsik A, Tábi T et al (2018) Compressive characteristics and low frequency damping of aluminium matrix syntactic foams. Mater Sci Eng A 739(2):140–148
8. Movahedi N, Murch GE, Belova IV et al (2019) Functionally graded metal syntactic foam: fabrication and mechanical properties. Mater Des 168(15):107652
9. Anbuchezhiyan G, Mohan B, Sathianarayanan D (2017) Synthesis and characterization of hollow glass microspheres reinforced magnesium alloy matrix syntactic foam. J Alloys Compd 719:125–132
10. Movahedi N, Murch G, Belova I, Fiedler T (2019) Effect of heat treatment on the compressive behavior of Zinc alloy ZA27 syntactic foam. Materials 12(5):792
11. Mukai T, Miyoshi T, Nakano S, Somekawa H, Higashi K (2006) Compressive response closed cell aluminum foam thigh strain rate. Scripta Mater 54:533–537
12. Banhart J (2001) Characterisation and application of cellular metals and metal foams. Prog Mater Sci 46:559–632

Part V
Industrial Applications

Metal Foams with Ceramic Inserts for Security Applications

René Vogel, Claudia Drebenstedt, Stefan Szyniszewski, Florian Bittner, Teresa Fras and Ludovic Blanc

Abstract Protective performance and a low weight of armour plates are the most important objective in security applications. Metallic foams can provide a high level of protection without an extensive mass and were used as a part of various laminated armour systems in the past. Embedding small ceramic parts in metallic foam combines the hardness of ceramics with the flexibility of the lightweight cellular matrix. The material has an extreme cutting resistance against mechanical tools such as angle grinder, drill bits, or other cutting devices. In several testing campaigns, aluminium foams were combined with different internal layouts of stacked or integrated ceramics. The cellular core was foamed directly to steel cover sheets or was bonded to different steel alloy faceplates. In additional tests, blast wave propagation of the metallic foam ceramic composite (MFCC) panels was evaluated. Furthermore, several ballistic tests were performed and compared to state-of-the-art armour materials. It was possible to achieve better performance in this comparison to standard armoured steel at a lower overall weight.

Keywords Metal foams · Forced entry · Blast protection · Ballistic protection

R. Vogel (✉) · C. Drebenstedt
Fraunhofer Institute for Machine Tools and Forming Technology IWU, Chemnitz, Germany
e-mail: rene.vogel@iwu.fraunhofer.de

S. Szyniszewski
University of Surrey, Guildford, UK

F. Bittner
Fraunhofer Institute for Wood Research, Wilhelm-Klauditz-Institut WKI, Hannover, Germany

T. Fras · L. Blanc
Explosive and Ballistic Protection, French-German Research Institute of Saint-Louis (ISL), Saint-Louis, France

N. Dukhan (ed.), *Proceedings of the 11th International Conference on Porous Metals and Metallic Foams (MetFoam 2019)*, The Minerals, Metals & Materials Series, https://doi.org/10.1007/978-3-030-42798-6_13

Introduction

There is a strong drive in industry to develop materials that are more capable to resist forcible entry attacks. As power tool technology has evolved and become more widespread and commercially available, research and development of materials that are able to withstand attack from now easily attainable tools has lacked. The ideal solution is a material combination, which is able to withstand attacks from even the most powerful handheld power tools, whereas it is not significantly increasing the mass or limiting the structural applications in which it can be utilized. By the modification of this multi-material design, it is even possible to fit the requirements of other security applications like lightweight protection panels against blast-initiated shock waves and projectiles.

Material Design and Manufacturing

The material consists of metallic and ceramic ingredients and requires metallurgical treatment. Firstly, the aluminium powder is mixed with titanium dihydride, TiH_2 (foaming agent) utilizing a rotating impeller to ensure a uniform mixture (Fig. 1). After the mixing stage, the powder mixture is consolidated via cold compaction in a compressor and then extruded through an extrusion die resulting in dense rods of material, which are cut into smaller pieces. Next, ceramic spheres and compressed aluminium powder rods are stacked in an orthogonal, grillage pattern and enclosed in a steel box using spot welds. The structure is then heated in a furnace to ca. 760 °C (depending on the melting range of the used aluminium alloy) for

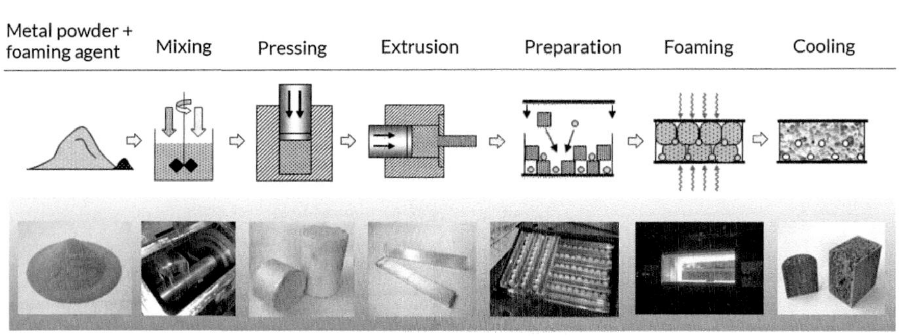

Fig. 1 Manufacturing steps. The metal powder is mixed with a foaming agent. Next, it is pressed into cylinders and extruded into diameter bars. The bars are stacked orthogonally and spaces between the bars are filled with 13 mm diameter, alumina spheres. Foaming agent decomposes at the temperature above 470 °C, while the aluminium powder melts, producing a metallic foam, which fills the space enclosed by the face sheets. The cellular metallic ceramic structure becomes stable after the foamed sample has been cooled below solidification range of the used alloy

between 15 and 20 min. The titanium dihydride begins to decompose at approximately 470 °C, releasing hydrogen gas. The release of the hydrogen expands the molten aluminium, creating voids until the cavity is filled completely. The components are subsequently cooled in calm air to produce stable cellular structures with embedded ceramic components.

The specimens were produced consisting of steel alloy (DC01) faceplates, cellular aluminium core (EN AW-6060) and an orthogonal layout of ceramic spheres and in some samples additional NiCr wires (Fig. 2) [1]. The spheres had 13 mm mean diameter, while aluminium foam had 2.1 mm average pore size only. The wall thickness is ranging from 30 μm to about 1000 μm. Thus, the ceramic inclusions were an order of magnitude larger than conventionally used ceramic particulate matter for reinforcement of metallic foams. The wires had an overall length of 100 mm and a diameter of 1 mm. In each sample, 30 wires were included. The ceramic spheres were sufficiently large to interact with small projectiles, drill or a cutting blade, and occupied 14% volume fraction of the cellular core. The aluminium foam matrix had density of 730 kg/m^3 and 73% porosity (air content), which ensured sufficient flexibility of the matrix. The overall metallic–ceramic hierarchical structure had 1140 kg/m^3 density. Sandwich panels with two 2 mm steel faceplates had 1780 kg/m^3 density in a 40-mm-thick panel configuration.

Cylindrical specimens were also manufactured to demonstrate that the technology can achieve high aspect ratio components such as beams or columns. The cylinders were also more suitable for computed tomography (CT) because the X-ray travel path through the foam was equal around the circumference, which reduced measuring artefacts.

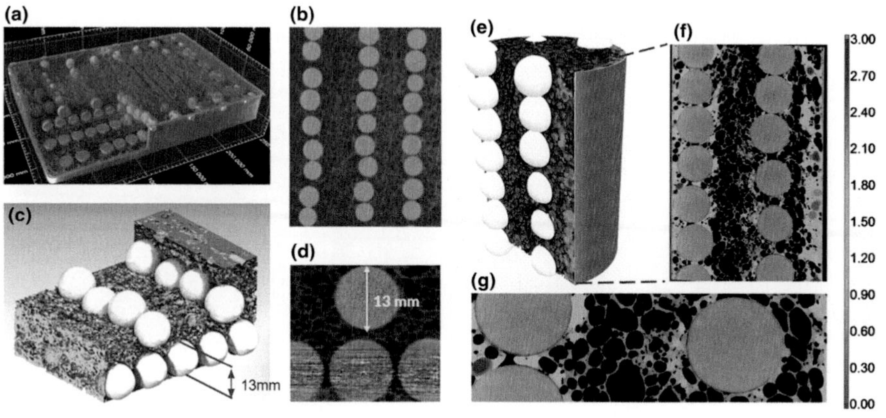

Fig. 2 Metallic cellular structure with embedded ceramic spheres. Images obtained from CT scans. **a** 245 mm × 172 mm × 40 mm sandwich panel. **b** Aluminium cells are an order of magnitude smaller than ceramic spheres. **c** Ceramic inclusions (white) are organized in an orthogonal pattern. **d** Ceramic spheres are not in contact with one another but are separated by aluminium cells. **e** Cylindrical specimens (60 mm diameter × 150 mm) had a vertical organization of ceramic spheres. Heatmap of aluminium foam wall thickness (mm). **f** Cross section of the cylinder with colours depicting cellular wall dimensions. **g** Thickness of aluminium cell walls varied mostly from 0.2 to 2.0 mm

Resistance Against Drill and Angle Grinder

The development of the material was driven by the aim to resist against extreme localized loads such as an angle grinder and power drill. The angle grinder attack represents the first load applied to penetrate the material. Different cutting discs with 115 and 125 mm diameter and a sapphire finish were used. The angle grinder revolved up to 180 revolutions per second, with rim velocity up to 80 m/s. The angle grinder achieved only a partial incision and itself experienced high wear. Its external diameter reduced from 125 to 44 mm after 60–65 s, and at that point, the cutting disc became ineffective (Fig. 3b–c). To provide a benchmark for our results, we tested one of the best currently available rolled-homogeneous armour (RHA) steels, distributed under the MARS 220 trademark.

The steel is quenched and tempered at temperatures below 200 °C resulting in a Brinell hardness of 440. The angle grinder completely penetrated the 10 mm MARS220 plate in 45 s (see Fig. 3a and Table 1), whereas the resistance mechanism exploited the contrast in stiffness between the ceramic spheres and the flexible aluminium cellular structure. Thereby, the energy of the rotating discs was partially transferred into the mechanical vibrations of the spinning disc and the embedded ceramic spheres integrated into the flexible cellular matrix. The tool operator experienced this as strong vibrations. It seems that the integration of the NiCr wires also results in a better resistance against the cutting tool.

In addition to the case where the cutting disc attacks the top of the sphere, other cases also were tested. During these tests, the angle grinder cuts in between two spheres and on the side of a sphere, halfway between the equator and sphere's pole.

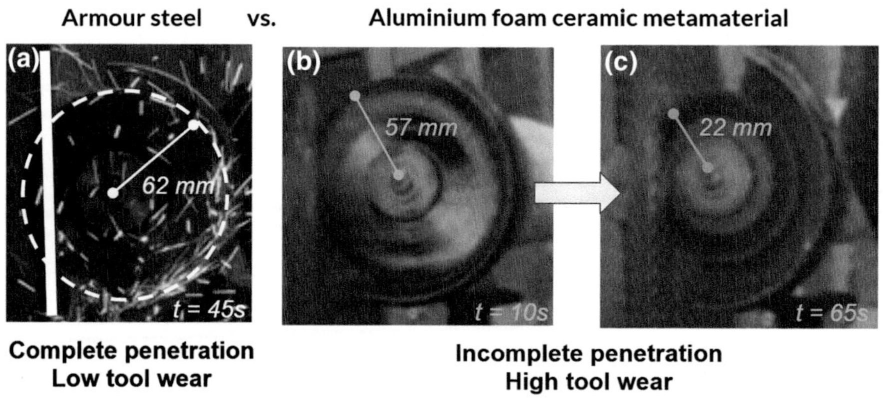

Fig. 3 Angle grinder attacks on conventional tank armour steel compared to our metallic foam ceramic metamorph. **a** The cutting disc achieved full penetration of 10 mm conventional armour steel plate within 45 s. **b** On the contrary, the disc experienced vibrations while cutting our architected metamorphic material and ceased to progress once the blade came in contact with the ceramic sphere. **c** The disc experienced high wear, its cutting interface eroded, and its diameter rapidly deteriorated rendering the disc inoperative in just over 1 min

Table 1 Angle grinder resistance test results

Test	Material	Duration (s)	Testing outcome	Penetration (mm)
A1	Al foam, NiCr wires ceramic spheres	62	Cutting blade completely worn	4–5
A2	Al foam, NiCr wires ceramic spheres	47	Cutting blade completely worn	5–6
A3	Al foam, NiCr wires ceramic spheres	57	Cutting blade completely worn	4–5
A4	Al foam ceramic spheres	78	Significant wear	6–7
A5	Al foam ceramic spheres	75	Significant wear	6–7
A6	Al foam ceramic spheres	65	Significant wear	7–8
A7	Mars 220	43	Low tool wear	Complete penetration
A8	Mars 220	46	Low tool wear	Complete penetration
A9	Mars 220	45	Low tool wear	Complete penetration

We observed the same outcome as the cutting disc diameter reduced rapidly in just over a minute and the angle grinder became also inoperative. This is because of the special sphere arrangement inside the sandwich. In each possible cutting scenario, the blade is touching at least one sphere. This is sufficient for the rapid tool wear.

In addition, we tested the resistance of our hierarchical material to a power drill. In all instances when a drill encountered a ceramic sphere, the drilling progress was stopped without any further progress (see Table 2). In these tests, the coated high-speed steel drill was not able to damage the ceramic anyway. The integration of NiCr

Table 2 Cordless drill resistance results

Test	Core material	Duration (s)	Testing outcome	Penetration (mm)
C1	Al foam, NiCr wires ceramic spheres	65	Little progress	7
C2	Al foam, NiCr wires ceramic spheres	63	Little progress	5
C3	Al foam, NiCr wires ceramic spheres	64	Little progress	7.5
C4	Al foam ceramic spheres	64	Significant progress	35
C5	Al foam ceramic spheres	30	Complete penetration of the panel	complete
C6	Al foam ceramic spheres	66	Significant progress	30

wires into the sandwich core results also in a better performance of the core material. The wires are an additional high-strength core element that could resist the drill or deflect it from the straight movement through the core.

Plastic Compressibility

In the first tests, compressed cylindrical samples (60 mm diameter × 150 mm height) were used to measure the mechanical properties of the material under quasi-static loading (Fig. 4a–b). Young's modulus, from loading–unloading tests, was E = 5.5

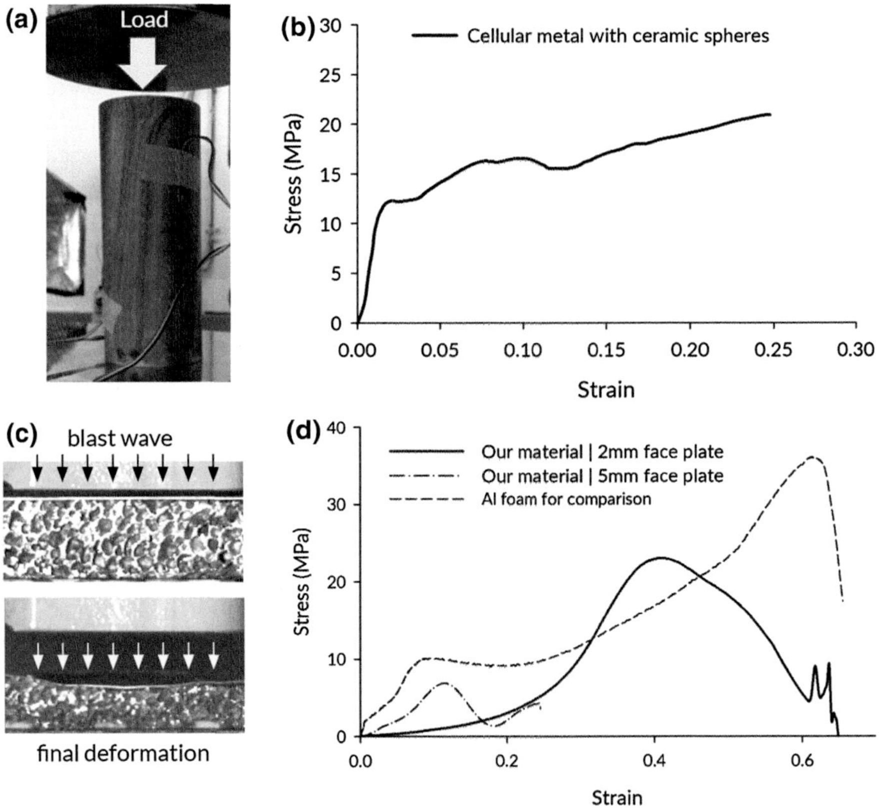

Fig. 4 Compressive testing. **a** Quasi-static compressive test of a cylinder sample. **b** Metallic foam with ceramic spheres showed distinct plateau and the ability for compressive deformation up to 25%. **c** Blast induced, dynamic compressive testing of a plate sample. **d** Material response showed a strong dependence on the imparted energy and strain rate of loading, which requires further studies. It is obvious that the classical aluminium foam is able to transform more kinetic energy in comparison to aluminium foam with metallic inserts. Because of the hard inserts, it is not possible to achieve a complete compression of the core material

GPa and the resulting yield stress was 8.1 MPa (according to ISO 13314 standard). The Poisson's ratio of the material was approximately 0.0 in the early stages of the loading and began ascending towards 0.5 from 10% engineering strain onward. The densification strain was about 0.25. The cellular structure showed significant deformability, exceeding 20% of engineering strain as expected from previous studies of cellular metals [2–6]. The inclusion of ceramic spheres did not have a significant effect under static loading because the deformations followed the paths of the least resistance via the cellular matrix in between the spheres.

For the testing under high dynamic conditions, 100 mm × 100 mm samples were produced. The shock wave was applied on one side of the sample. The planar shocks were generated with an explosive-driven shock tube (EDST) [7–9]. The shock tube had a square of 100 × 100 mm^2 external and 80 × 80 mm^2 internal section. The total length was 1750 mm. 50 g of C4 explosive charges were detonated 50 mm in front of the tube, generating an incident load with a pressure of 30 bars, an impulse of 9.33 bars-ms, and a phase duration of 1.32 ms at the far end of the tube. These effects are equivalent to those of 567 kg of TNT placed approximately 4.72 m away from the target. The plate sample with 25 mm thickness of the metallic foam ceramic material was placed at the end of the tube, against a rigid chassis. Behind the rearward support, a load sensor was installed. Due to the explosive shock tube setup, it was possible to record the compression process with a high-speed camera without the effects of the fireball.

Firstly, we tested the material with 2 mm faceplate (Fig. 4c). The faceplate velocity increased up to 70 m/s, giving a top strain rate of 2800 1/s, preceded by acceleration at the start and deceleration at the end of the deformations. Given the blast impulse, we were able to estimate the initial velocity of the front plate and consequently its kinetic energy. The energy imparted by the blast was 2087 J and 150 J was absorbed up to 20% strain, and additional 650 J during the subsequent hardening (Fig. 4d). Ceramic spheres impacted the back plate at around 60% strain, which corresponds to the oscillations at the end of the stress–strain curve in Fig. 4d. In total, 38% of the blast energy was absorbed.

Even though the material was highly compressible, its stress–strain curve was atypical compared to a classical aluminium foam without ceramic inserts and a density of 616 kg/m^3. Metallic foam with ceramic spheres had low resistance up to 20% strain, and noticeable strain rate hardening up to 40% strain. The deformed shape of the face sheet was highly non-uniform (see Fig. 4c), which indicated that the core was more compliant in between the spheres. Another configuration, containing thicker 5 mm faceplates, also was tested. Enlarging the weight of the plate reduced the energy imparted into the material to 834 J due to the increased inertia of the faceplate. The maximum strain rate was lowered to 1600 1/s. Approximately, 29% of that energy (240 J) was absorbed before the densification in the second case, and the metallic foam ceramic core was only partially compressed.

Interestingly, the dynamic response resembled the quasi-static material curve more closely at the lower strain rate. The qualitative difference between the static and dynamic stress–strain curves may arise from the complex refractions and reflections of the shock wave encountering ceramic spheres in a compliant metallic foam. Le et al. demonstrated computationally that spatial hierarchies made of materials with contrasting stiffness and density can disperse, focus, or even change the direction of the blast wave propagation [10]. Overall, the metallic foam ceramic structure is highly sensitive to the strain rate of the loading. The hardening of the material was not ideal from the perspective of a blast engineer because it may allow for the transmission of high pressure to the protected structure. However, our two tests demonstrated that the energy absorption and peak-transferred stress can be controlled by varying the faceplate thickness. Further optimization of the volume fraction of ceramic spheres, their topology, the thickness of cellular metal, and its density to maximize the blast mitigation properties are beyond the scope of this study.

Ballistic Testing

For ballistic testing, different sandwich configurations were designed and tested at ISL laboratories. The testing was done according to VPAM standards for the levels 9–11, starting at the highest level. In this case, a sample that was penetrated by a higher energy projectile could be reused for testing at lower VPAM standards (Table 3).

During the tests, it was possible to fit VPAM level 9 and 10. In samples with two layers of ceramic spheres, the bullets passed the impact side cover sheets of the sandwiches. Subsequently, the projectiles were shattered by the ceramics inside the metal foam. Finally, the fragments of bullets and ceramics were stopped at the Mars 240 back plate (see Fig. 5). The standard configuration for stopping a Dragunov projectile (VPAM level 10) is an 18-mm-thick armoured steel plate with an areal mass of 141 kg/m^2. The areal mass of the successful metal foam configuration 1.1 is 113 kg/m^2, 20% lighter in this case.

Bullets with hard metal core were also shattered inside the sandwich, but the fragments perforated the back plate. Obviously, the energy dissipation inside the core is not sufficient in this case (see Fig. 6). Future work is necessary to test other sandwich configurations to meet higher level of ballistic threats. Further investigations to reduce weight and thickness of the panels are also necessary.

Summary

This work is part of a larger effort to build up know-how about architected materials on metal foam basis for security applications. By integration and combination of metal cellular structures, ceramics, and ballistic steel, it is possible to build up a multifunctional platform, which combines ductility with hardness. Not only it is possible to withstand high-performance cutting tools, but it is also possible to integrate high compressive ductility and energy dissipation under blast. Furthermore, it is possible to meet VPAM levels up to 10 at a lower areal weight of the panel in comparison to standard steel platings. In future, it will be possible to enhance the ballistic resistance

Table 3 Applied ballistic testing standards

7.62 mm × 51 AP P80 0.308 Win FMJ/PB/HC	7.62 mm × 54 API B32 Dragunov FMJ/PB/HC	7.62 mm × 51 AP W C .308 Win FMJ/PB/WC
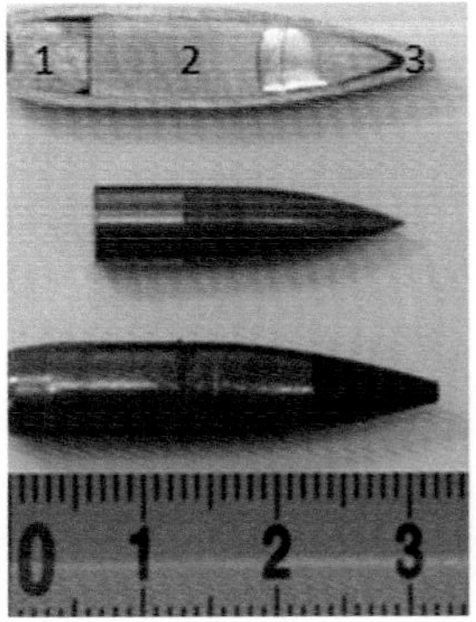	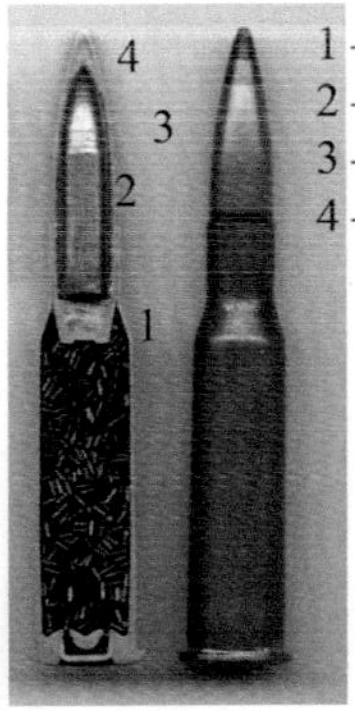	
Impact energy: 3261 J Impact velocity: 820 ±10 m/s	Impact energy: 3846 J Impact velocity: 860 ±10 m/s	Impact energy: 3633 J Impact velocity: 930 ±10 m/s
VPAM level 9	VPAM level 10	VPAM level 11

Fig. 5 VPAM 10 ballistic testing configuration and results. The bullet was fragmented and stopped by the steel back plate, 8 mm MARS 240, Configuration. 1.1

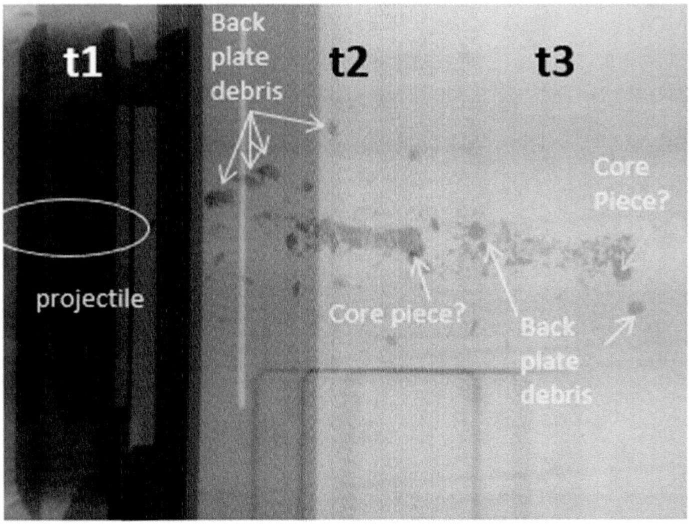

Fig. 6 VPAM 11 ballistic testing results. The bullet was fragmented and penetrated the steel back plate, 8 mm MARS 190, the X-ray image shows 3 time steps of the test: t1—incoming bullet inside the sample, t2 and t3—shattered projectile, ceramics, and back plate after penetration of the back plate

of light armour plates or to achieve higher resistance levels at 20% lower areal weight. Blast resistance, moderate ballistic protection, and non-cuttable features are combined advantages that could be useful in many infrastructural applications. However, further optimization and weight reduction could also enable the possibility for usage in armoured vehicles or other mobile uses.

While the future possibilities of this technology were shown, further investigations are necessary to build up the right architecture for each application. Yet we know that, for example, internal contacts, material hardening, ceramic fracturing, and the projectile movement have an effect on the quality of the multi-threat protection. But it is essential to know which internal processes, alone or in combination, are most significant. To achieve this goal, additional tests with updated material combinations will be tested soon at the French-German Research Institute of Saint-Louis (ISL).

References

1. Neugebauer R, Lies C et al (2007) Adhesion in sandwiches with aluminum foam core. Prod Eng 1(3):271–278
2. Gibson LJ, Ashby MF, Ashby M (1999) Cellular solids: structure and properties. Cambridge University Press
3. Ashby M (2000) Metal foams: a design guide. Butterworth-Heinemann
4. Ashby MF (2011) Materials selection in mechanical design. Butterworth-Heinemann
5. Smith BH, Szyniszewski S, Hajjar JF, Schafer BW, Arwade SR (2012) Characterization of steel foams for structural components. Metals 2:399–410
6. Smith BH, Szyniszewski S, Hajjar JF, Schafer BW, Arwade SR (2012) Steel foam for structures: a review of applications, manufacturing and material properties. J Constr Steel Res 71:1–10
7. Blanc L, Sturtzer M-O, Schunck T, Eckenfels D, Legendre J-F (2018) Blast wave mitigation using multi-phase solid material in a sandwich cladding, Ref: PU-271/2018
8. Ousji H et al (2017) Air-blast response of sacrificial cladding using low density foams: experimental and analytical approach. Int J Mech Sci 128–129:459–474
9. Reddy CJ, Madhu V (2017) Dynamic behaviour of foams and sandwich panels under shock wave loading. Procedia Eng 173:1627–1634
10. Le C, Bruns TE, Tortorelli DA (2012) Material microstructure optimization for linear elastodynamic energy wave management. J Mech Phys Solids 60:351–378

The Challenge of Open Cellular Metal Foam Production

Christian Hannemann, Mandy Uhlig, Thomas Hipke and Iris Meier

Abstract Five years ago Fraunhofer-IWU started research in open cellular metal structures as the potential applications increased based upon their exceptional properties. Heat exchangers, filters, and battery electrodes are examples of potential applications. As casted aluminum foams were in serial production and Fraunhofer in close cooperation with the manufacturer, the interest was limited to prototyping of individual foams especially in combination with solid regions for fixing, encasing, or sub-division. New parameter combinations and technological processes were developed and the advantages, limitations, and potentials are described within this document. Alternatively, galvanic coating of PUR foams with copper, nickel, and other metals is common, again buyable and state of the art. R&D activities related to application development and adapted geometries not yet realizable by conventional mass production will be shown. Investigations for graded strut-structures and cylindrical tube-like parts were intensified to reach serial production level. The lecture represents investigations, results, dropped, and new approaches of all routes.

Keywords Foam · Open cellular · Salt · Mold material

Background

About 5 years ago, Fraunhofer-IWU recognized an increasing demand on open cellular metal foams and lost the partner collaborating in that subject. Several production technologies have been investigated to find the best suitable solution for customer

---

C. Hannemann (✉) · M. Uhlig · T. Hipke · I. Meier
Fraunhofer-Institute for Machine Tools and Forming Technology, Reichenhainer Str. 88, 09126 Chemnitz, Germany
e-mail: christian.hannemann@iwu.fraunhofer.de

M. Uhlig
e-mail: mandy.uhlig@iwu.fraunhofer.de

T. Hipke
e-mail: thomas.hipke@iwu.fraunhofer.de

N. Dukhan (ed.), *Proceedings of the 11th International Conference on Porous Metals and Metallic Foams (MetFoam 2019)*, The Minerals, Metals & Materials Series, https://doi.org/10.1007/978-3-030-42798-6_14

applications and future fields of research. The market for these structures grows continuously especially because of the large inner surface. Heat exchangers, filter applications, and battery electrodes are just some examples of the potential applications. As there have been casted aluminum foams available on the market, Fraunhofer-IWU cooperated with m-pore GmbH in Dresden whenever asked for open cellular aluminum. Worldwide there are only view manufacturers such as ERG Aerospace Corporation in the US. After the partner m-pore was taken over by Mayser GmbH & Co. KG, the production was temporarily stopped provoking own activities of the Fraunhofer-IWU to establish a small prototypical production line for open cellular structures and that way fulfill the customer demand. The technology used is identical to the ERG or m-pore process and well described in literature [1, 6] and shown in Fig. 1.

Challenge

Soon it was recognized that the technology for a serial production is far away from being competitive if produced at an institute. An analysis of all relevant production steps, engaged materials, and personal costs showed a maximum effect caused by the mold material and as predicted the personal costs. As most of the production steps are manual and hard to be automated, new approaches were needed. The internal investigations of Fraunhofer-IWU resulted in a patent application in 2015 being rejected because of similar attempts of the Fraunhofer-IFAM. A disclosure document had been published shortly after the mentioned patent application [3]. Even with a slightly different background, objective similarities were obvious. Anyway, it was found that the conventional method is capable of a maximum production of about 1.000 m^2/a.

The intention of technology modification is based upon a high-pressure die casting process in combination with an investment casting core. As the process is aiming at identical final products, it starts with the same master pattern out of reticulated polyurethane foam cut into the desired shape. The structure needs to be modified and the cross shape of the struts enlarged by an automated wax coating and is then enclosed into a ceramic plaster to create the mentioned core.

During the followed heat treatment, the mold is burned to its final condition parallel to the elimination of the master pattern that is leaving a negative geometry as hollow channels in the mold material. The heat treatment ends at a mold temperature ready for the casting. One major difference of the conventional investment casting technology is that the mold block shows a dimension just as the master pattern (Fig. 2). That way all struts within the outer surface cause openings for the melt to infiltrate the block during the followed casting. The mold block is placed in the cavity of a pressure die casting mold that also contains the gating system. Afterward, the mold is closed and filled by a low-velocity casting process. Evacuation of the mold can support the filling as described in the Vacural process (Foundry T&E [4]) and improve the infiltration. As common in the die casting the gate system can be used

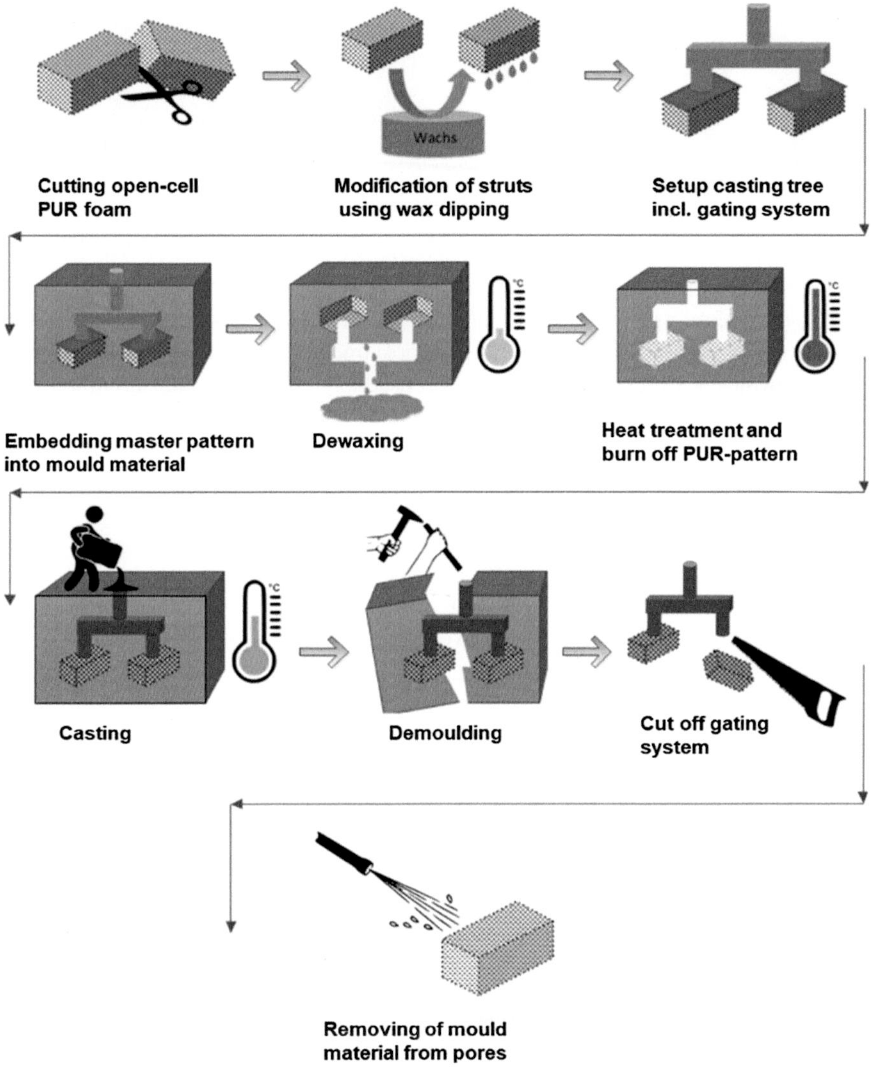

Fig. 1 Technological process of manufacturing open cellular aluminum foam

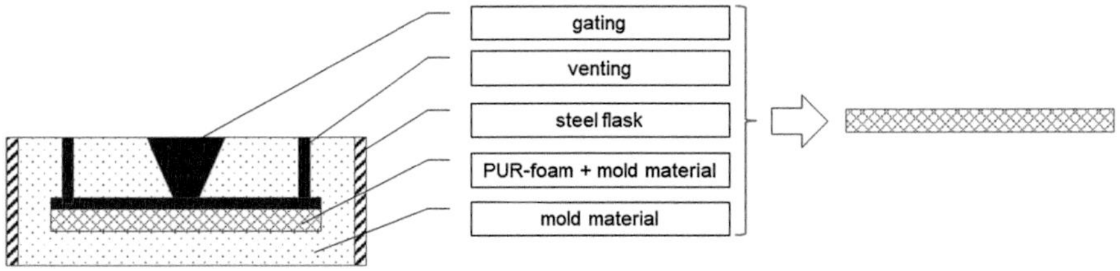

Fig. 2 Reduction of the required molding material to minimum size (conventional mold-left, approach-right)

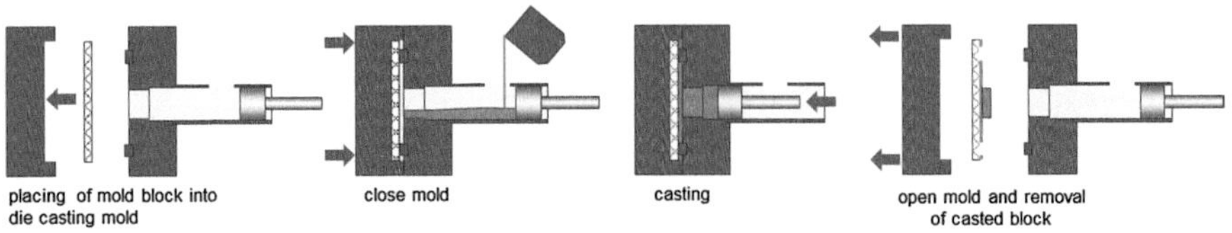

Fig. 3 New approach as technological process of manufacturing open cellular aluminum foam

for robot handling to remove the casted part out of the die casting mold and finally to remove the investment molding material from the structure. This process could reduce the costs of the manufacturing to approximately 20% of the current state of the art. Further advantages are increased reproducibility and decreased amount of needed mold material to about one-third of the conventional technology (Fig. 3).

As common in die casting, the gate system is used for robot handling to unmold and during the final removal of the investment mold material by high-pressure water jet.

The described process could reduce the costs of the manufacturing to approximately 20% of the current state of the art. There is a higher reproducibility and the energy efficiency of the manufacturing increases. In addition, the invested mold material is reduced to a third of the conventional process.

Current Production Status

The company Mayser stopped production of the open porous aluminum foam in the end of 2018. That decision was caused by missing customer interest (based upon the prices achievable with the used technology) and decreased personal capacities. Currently there are large amounts of semi-finished foams available on stock and the Fraunhofer-IWU is qualifying a new producer to establish an open cellular aluminum foam production. Aside from the costs for the foam production, customers have been questioned for application hesitations. One critical fact often mentioned relates to the tolerances in reproducibility caused by the chemical foaming process of the PUR being the major parameter for the predictability.

Approaches

Additive manufacturing is a booming technology that reaches almost all technical fields from medical engineering, to automotive and even food industry. The biggest advantage is the flexibility and freedom of geometry. In technical applications, 3D printing is well established and state of the art. The largest potential in open cellular metal structures relates directly to the problem described above as it is possible to design the cellular structure directly for the need of the application.

At the Fraunhofer-IWU, 3D wax printing has been used to realize cellular wax structures as master pattern for investment casting. The almost unlimited freedom of geometry resulted in many different constructions adapted to the future application (Table 1).

All of the shown structures show the behavior of casted structures as they are. There is no layer or staircase effect recognizable as the resolution of the printer is 16 μm.

In a project funded by the German Federal Ministry for Economic Affairs and Energy, the heat transition inside the foam structure has been investigated. Especially the effect of adapted geometry was in focus and simulated as well as practically verified (Fig. 4).

During the project, lattice structures have been investigated with graded and homogeneous strut-cross shape and the results are shown in Fig. 5.

At the moment, there are funding proposals running to the further development of these structures aimed at a large serial production technology without additive

Table 1 Samples for cellular structures realized by 3D printing in combination with investment casting

Name	Pentamode structure	Schwarz D-structure	Schwarz P-structure	Graded structure (Kelvin cells)
Picture	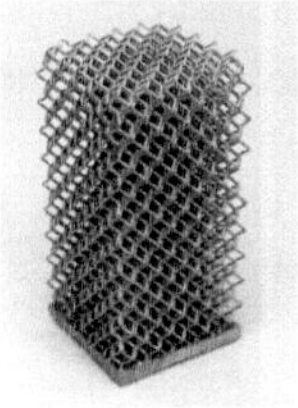			
Specific parameter	Strut structure specified to show bending only in intersections	Minimal surface structure for chemical functionalization	Two divided interlaced volumes for perfusion	Graded struts from introduction plate reduced to half size cross shape
Application	Mechanical sensor	Chemical water treatment	Heat storage	Heat exchange

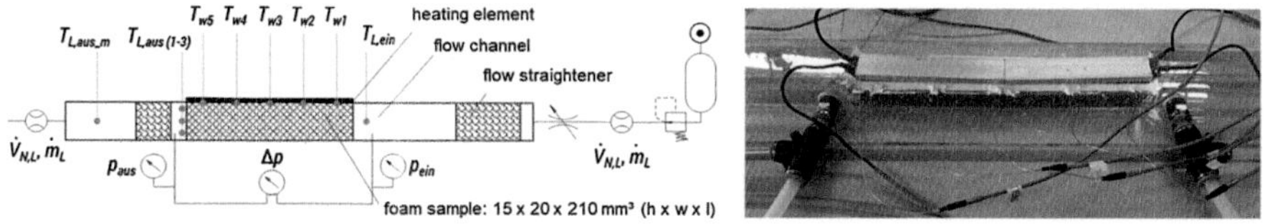

Fig. 4 Analysis of pressure loss and heat transfer

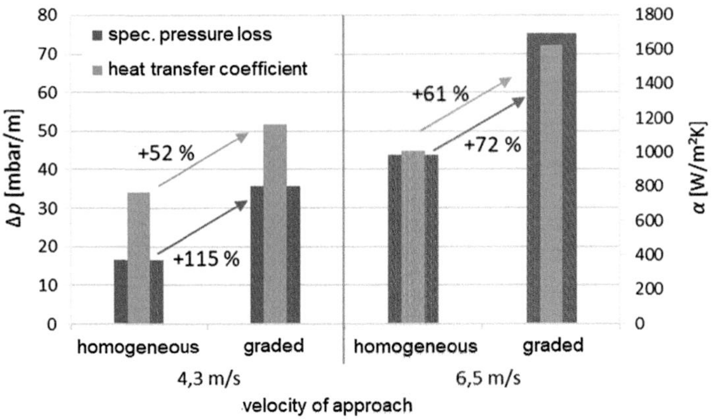

Fig. 5 Effect of graded strut geometry

manufacturing. That way the application development chain contains a prototyping stage realized by 3D-printed wax pattern followed by a serial production of the master pattern in wax injection of the cellular structures finally in both cases resulting in casted parts.

Another approach followed to establish and expand the open cellular aluminum structures is dealing with the major cost factor: mold material. A specialized development of mold material to fulfill all requirements for the open cellular foam production will be part of a national project between the Fraunhofer-IWU and the Hochschule Pforzheim. The focus will be at resource efficiency and recyclability as well as ways to unmold the casted structures without mechanical deformation.

Especially the last argument is part of another approach that is currently under investigation. There have been several customer demands for foam-filled pipes meaning a tube-like structure that shows an open cellular structure inside. That way fluids or gases can permeate the structure and react with the large surface of the foam or exchange heat between the fluid and the enclosing shell also using the struts to transfer the heat via the large contact surface. Conventional technologies are limited in depth to about 100 mm as the mold material is removed mechanically by high-pressure water jet. Even supporting that process by chemical reactive additives showed minor success. The part shown in Fig. 6 is casted in one piece including the outer shell. That way the connection between cellular structure and shell is perfect showing a perfect heat transfer.

With a length of 160 mm, a cross shape of 7 × 45 mm, and a wall thickness of 1 mm, it stands for the reachable maximum in the conventional technology. The sidewalls are meant to induce the heat into the system and shall be minimized but still leakproof even with pressurized media running through. As the customer asked for longer parts and an outer shell of 0,5 mm thickness, the order had to be refused creating new demands for investigations.

It is common practice in the casting industry to work with cores to realize inner geometries. Well known are salt cores in sand and even die casting [5, 8]. As salt would

Fig. 6 Heat exchange structure all cast in one piece

be dissolvable in water and that way could be eliminated out of the mentioned tube-like structures, the idea of using salt as mold material occurred. The first fundamental tests show promising results even with conventional table salt.

The salt is ground and sieved into defined fracture sizes and several mold samples have been realized. As known, salt is hygroscopic and tends to agglutinate. This effect was used by moistening the slightly compacted salt crystals. After a drying procedure depending on the degree of moisture, a solid mold block resulted. It was found that with variation of the grain size the surface roughness could be tailored from sand to investment casting quality (Fig. 7).

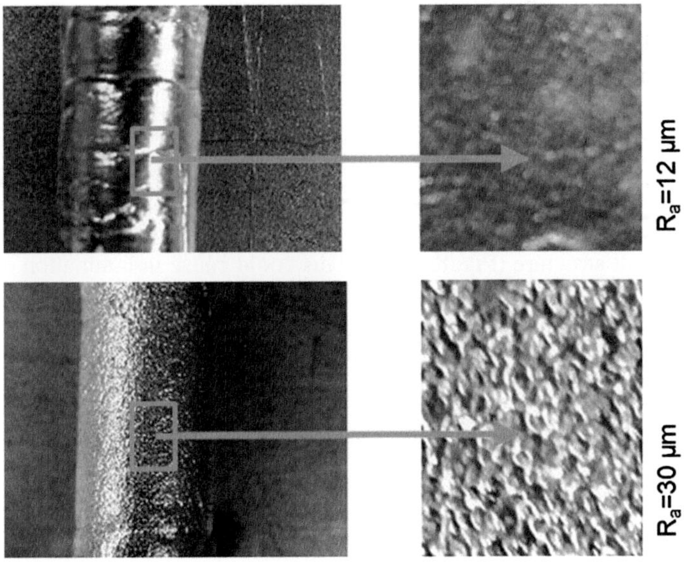

Fig. 7 Casted samples using different salt fractures

Fig. 8 Samples of compacted salt (left: three-point bending test; right: salt-mold wax infiltrated)

With the applied NaCl there have been no chemical reactions in casting with neither the tested tin nor aluminum. The feasibility of a partly mold material substitution of the conventional plaster/ceramic type by salt for the manufacturing of open porous structures is currently under investigation. To realize the infiltration of the salt particles into the PUR structure, a fluid bed of salt followed by small vibration compaction is used [2, 10]. Especially for the drying of salt these fluid beds are common state of the art and in large serial use [7]. Some samples of the compacted salt for property analysis are shown in Fig. 8.

The approach solves further problems described before. Once the salt is removed by dissolution it can be recovered by water evaporation or another drying process followed by new fracturing. That way the material costs are reduced to a minimum and the environmental footprint is very limited.

The dissolving process happens within seconds and can even be supported by ultrasound vibration. In the current molds, there have been no residuals left in the top-level surface of the casted pieces or in undercuts. To control about remaining salt in the casted structure, the water used for the dissolving can be tested by salt meters available on the market recognizing even minimal amounts of ions in the water (e.g., PCE Deutschland GmbH [9].

There are several problems to be considered, such as the infiltration of model wax into the salt structure (Fig. 8; right). This could cause inflammation during the dewaxing process or while increasing the temperature for the casting and that way easily extending above the melting temperature of the salt. Also the wetting of the melt at the salt crystals seems to be excellent which can cause negative effects of infiltration into the mold structure as well as positive effects of a high surface quality depending on the grain size and density of the salt mold.

Call for Cooperation

As described above, some of the approaches to regain someone into the producer's role have already started. Anyway, the topic and customers are pushing the development and we are looking for partners/projects/funding to follow and further develop all of the shown approaches.

References

1. Ashby MF, Evans T, Fleck NA, Hutchinson JW, Wadley HNG, Gibson LJ (2000) Metal foams: a design guide. Elsevier
2. Corportaion K (2019) Fluid bed dryers, coolers and moisturizers
3. Eilers J, Haesche M, Heuser M, Wöstmann F-J (2013) Verfahren zur Herstellung komplex geformter Gussteile mit einer Druckgießmaschine. DE102012006572A1
4. Foundry T&E (2019) Vacural-Verfahren [WWW Document]. Gießerei-Lexikon. URL https://www.giessereilexikon.com/giesserei-lexikon/Encyclopedia/show/1018/?L=28&cHash=fa70abc99c24ba84803fb708ad2266a7. Accessed 31 July 2019
5. Fuchs B, Eibisch H, Körner C (2013) Core viability simulation for salt core technology in high-pressure die casting. Inter Metalcast 7:39–45. https://doi.org/10.1007/BF03355557
6. Hipke T, Lange G, Poss R (2007) Taschenbuch für Aluminiumschäume, 1, Aufl edn. Alu Media, Düsseldorf
7. Imtech Ventilex (2019) fluid-bed-salt-drying-.pdf [WWW Document]. URL http://www.ventilex.com/wp-content/uploads/2014/10/Fluid-Bed-Salt-Drying-.pdf. Accessed 6 Aug 2019
8. Kallien L (2016) Salzkerne im Druckguss 32–43
9. PCE Deutschland GmbH, 2019. Products for "salt"| PCE Instruments [WWW Document]. URL https://www.pce-instruments.com/us/?action=Query&-query.&query.stichwort=salt. Accessed 6 Aug 2019
10. Raihane A, Bonnefoy O, Chaix J-M, Gelet J-L, Thomas G (2011) Analysis of the densification of a vibrated sand packing. Powder Technol 208:289–295. https://doi.org/10.1016/j.powtec.2010.08.018

Part VI
Metal Foams in Thermal Systems

Investigation of Heat Transfer Performance of PCM/Copper Foam Composite Heat Sink for Electronic Devices

Ratiba Sabrina Ferfera, Brahim Madani and Rafik Serhane

Abstract This experimental and numerical study is carried out in order to study the thermal performance of a composite plate (paraffin/copper foam) intended to cool portable electronic devices. In order to simulate the transport phenomena inside the plate, a microcellular 3D model with cubic centered body shape for a single cell is developed using COMSOL Multiphysics software. To calibrate the numerical model, experimental transient temperature profiles are used as inputs. The effects of the composite insert on heat dissipation are investigated and compared to other types of heat sinks.

Keywords Heat transfer · Phase change material · Metallic foam · Experimental · Numerical simulation

Nomenclature

a	Unit cube side, m
C_p	Specific heat capacity, J kg^{-1} K^{-1}
d_p	Pore diameter, m
H	Latent heat, kJ kg^{-1}
h	Free convective coefficient, W m^{-2} K^{-1}
k	Thermal conductivity, W/m K
T	Temperature, K
T_{m1}	PCM melting area lower temperature, K

R. S. Ferfera · B. Madani (✉)
Laboratory of Multiphase Transport and Porous Media (LTPMP), Faculty of Mechanical and Process Engineering (FGMGP), University of Sciences and Technology Houari Boumediene (USTHB), 16111 El Alia Algiers, Algeria
e-mail: bmadani@usthb.dz

R. Serhane
Division Microélectronique et Nanotechnologie (DMN), Centre de Développement des Technologies Avancées (CDTA), City August 20th, BP 17, 1956 Baba Hassen, 16303 Algiers, Algeria

N. Dukhan (ed.), *Proceedings of the 11th International Conference on Porous Metals and Metallic Foams (MetFoam 2019)*, The Minerals, Metals & Materials Series, https://doi.org/10.1007/978-3-030-42798-6_15

T_{m2} PCM melting area upper temperature, K
t Time, s
x, y, z Cartesian axis direction, m

Abbreviations

PCM Phase Change Material
MF Metallic Foam
BCC Body-Centered Cubic

Greek Symbols

β Thermal expansion, 1 K^{-1}
ε Porosity
θ Liquid fraction
μ Dynamic viscosity, Pa s
ρ Density, kg m^{-3}
ω Pore density, PPI

Index and Exponent

0 Initial—ambient
l Liquid
s Solid

Introduction

Due to the rapid development of the economy, industry, and human activity in general, energy consumption has increased significantly in recent years. Obviously, thermal energy storage can reduce the mismatch of the energy supply and energy demand [1]. Latent heat storage through the use of phase change material (PCM) is regarded as one of the most effective and attractive thermal energy storage methods [2]. This is due to its large heat storage capacity as it involves the fusion latent heat and a nearly constant temperature during the phase change process. Phase change materials (PCMs) are widely utilized in many applications, such as in building industries [3] and solar thermal application [4]. Also, note that there are many recent applications of PCMs for the thermal management of portable electronic devices [5, 6].

However, PCMs present a major disadvantage, their very low thermal conductivity, which causes poor heat diffusion of the heat sinks. Many researches are carried out to improve the thermal conductivity of PCMs. Therefore, many solutions have been adopted, as addition of graphite powder [7], carbon fibers and nanocarbons [8], fins [9] or metallic foams [10, 11].

Open cell metallic foams are known to have many interesting thermal and mechanical properties: low density, high permeability, and high thermal conductivity [12]. For these reasons, the PCM/metal foam (PCM/MF) composite is promising in electronic devices' thermal management application.

The aim of this work is to determine experimentally the effects of adding PCM/MF composite material on heat sink's thermal performances. In this case, a paraffin with melting point from 42 to 48 °C and latent heat of 160 W/m K is chosen as the PCM to impregnate in a copper foam with a porosity and a pore density of 96.1% and 10 PPI, respectively. The obtained results are compared to heat sinks equipped with pure paraffin and an empty copper foam. A microcellular model like a body-centered cubic geometry (BCC) is developed using the COMSOL Multiphysics Software. The wall temperature which has been obtained experimentally is used as a boundary condition in CFD simulation.

Experimental Procedure

Experimental Setup

Figure 1 shows the schematic of the setup to study the heat transfer performances of the heat sink equipped with the copper metallic foam (with porosity of 96.1% and pore density of 10 PPI which is provided by POROMETAL®) infiltrated with paraffin wax as a PCM (with latent heat of 160 kJ/kg K and melting temperature from 42 to 48 °C). The mass of paraffin used in this experiment for the two cases of pure paraffin and PCM/Cu foam composite are 17 g and 7.85 g, respectively. The properties of copper and PCM are listed in Table 1. Regarding the apparatus, it consists of a Plexiglas box, for containing the composite sample of $80 \times 50 \times 5$ mm^3 as volume and the heat source plate. An empty space of 10 mm is left between the upper surface of the sample and the box to prevent the volume expansion of the melted paraffin. Three K-type thermocouples are used to measure the transient temperature of the PCM/MF composite sample and the heater; their positions are shown in Fig. 1. The transient temperature data are recorded using data acquisition system (DAQ) and stored in computer memory for analysis.

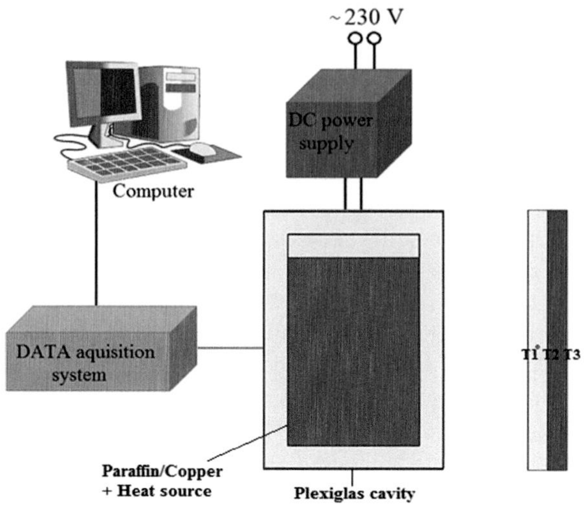

Fig. 1 Schematic illustration of the experimental apparatus in the left and thermocouple's position in the right

Table 1 Paraffin and copper properties

	ρ (kg m^{-3})	K (W m^{-1} K^{-1})	C_p (kJ kg^{-1} K^{-1})	T_{m1}, T_{m2} (°C)	H (kJ kg^{-1})
Paraffin	880	0.2	2000	42–48	160
Copper	8920	380	380		–

Experimental Results and Discussion

Figure 2 compares the transient temperature of the heat source for several cases: isolated and bonded to three different heat sink plates (copper foam, pure paraffin and composite PCM/MF), for a duration of 8000 s heating process. It is shown that

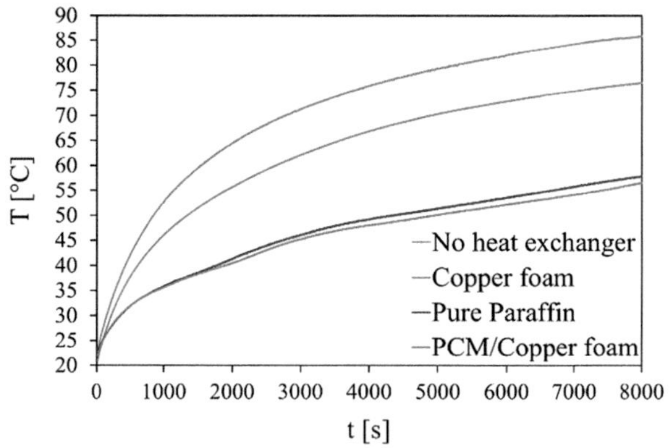

Fig. 2 Heat source transient temperature

the temperature of the heat source without the heat sink plates increases faster than that equipped with heat sink plates, reaching temperature of 85 °C during charging process. The copper foam insert drops the heat source overheating by 6.4 °C compared to the isolated heat source case. Unfortunately, the heat source with empty copper foam reaches the temperature of 75 °C, which is still too high for electronic components. For pure paraffin and PCM/MF composite as heat sinks, the overheating of heat source dropped by 22 °C and 23.5 °C, respectively. Note that the fatidic temperature of 60 °C during charging process is not reached in these two cases. The rate of the heat dissipation improvement reached 34%. Regarding the heat dissipation and the overheating prevention of the heat source, the figure shows insignificant difference whether pure paraffin or PCM/MF is used as a heat sink. However, the actual effects of the PCM/MF insert are given in the section below.

The difference in temperature between the heat source and heat sinks at different locations of thermocouples is shown in Fig. 3. ΔT_{1-2} and ΔT_{1-3} refer to the difference in temperature between the heat source (thermocouple T1) on one hand and the two facets of the samples on the other hand: the heated one (thermocouple T2) and the opposite one (thermocouple T3). For pure paraffin, ΔT_{1-2} and ΔT_{1-3} are of 1.9 °C and 9.9 °C, respectively. ΔT_{1-3} for pure paraffin is highest due to the paraffin's low thermal conductivity, which prevents the heat diffusion through the plate heat sink. For ΔT_{1-2} and ΔT_{1-3} with PCM/MF composite, they are equal to 1.1 °C and 3.8 °C, respectively. The PCM/MF composite improves greatly the heat dissipation through the heat exchanger by about 62%. This is due to the improvement of the thermal conductivity by copper in the composite material insert.

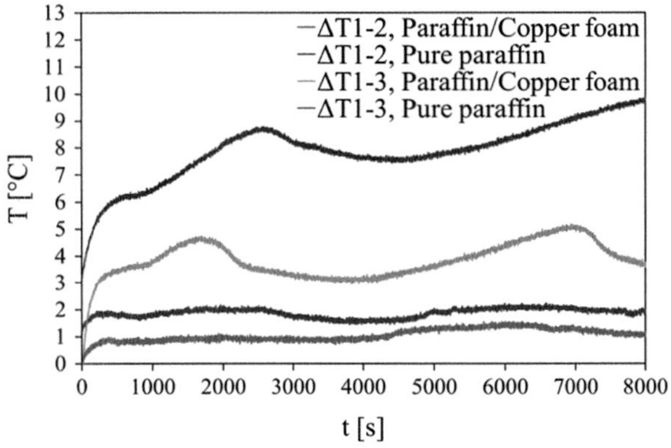

Fig. 3 Temperature difference between the heat source and heat sinks at different locations of thermocouples

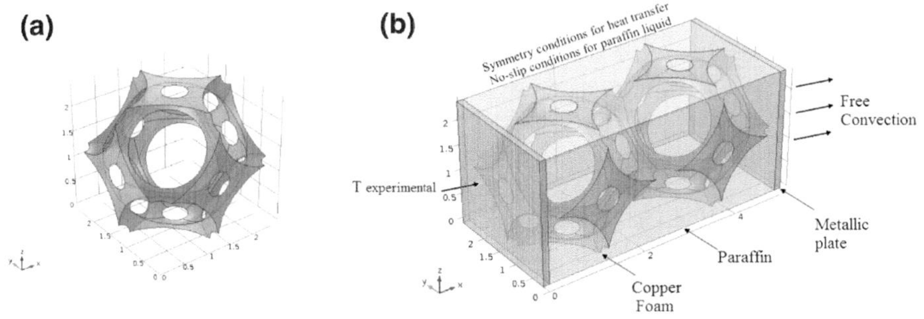

Fig. 4 Schematic representation of **a** BCC shape **b** 3D model of paraffin impregnated in a copper foam

Numerical Simulation

Geometric Configuration and Boundary Conditions

In this work, the metallic foam cell is considered having a body-centered cubic (BCC) shape. In order to obtain the BCC shape for the idealized cell, it is assumed that the pore diameter is spherical, and an eighth sphere (diameter of d) is subtracted from each corner of a unit cube (side of a) and a sphere from the center of the unit cube, as shown in Fig. 4a. To have an open unit cell, the sphere diameter must be greater than the unit cube side.

The geometry of the 3D model is composed of two unit cells with body-centered cubic (BCC) shape for each one. The cells are made from copper foam incorporated with paraffin, as shown in Fig. 4b. Two copper plates are added on both faces of the composite. The heated wall in the left side is submitted to the transient temperatures profile which is obtained experimentally. To take into account the heat losses, the opposite wall is supposed to be exposed to a free convection stream with a constant convective coefficient of $h_0 = 2.5\ \text{W/m}^2\ \text{K}$ at room conditions of pressure and temperature. Due to the symmetry and direction of the heat flux in the x-axis, the sidewalls are assumed to be adiabatic and the non-slip condition is applied at box walls in contact with the paraffin liquid phase.

Governing Equations

The governing equations in this heat transfer study include liquid paraffin continuity equation (see Eq. 1), liquid paraffin momentum equation (see Eq. 2), thermal conduction and convection in paraffin (see Eq. 3), and thermal conduction in copper ligaments (see Eq. 4).

$$\rho_l \frac{\partial U}{\partial t} + \rho_l (U \cdot \nabla) U = 0 \tag{1}$$

$$\rho_l \frac{\partial U}{\partial t} + \rho_l (U \cdot \nabla) U = -\nabla.[p + \mu(\nabla U)] + \rho_l g \alpha (T - T_0) \tag{2}$$

$$\rho_l \left(C_{P_f} + H \frac{d\theta}{dT} \right) \frac{\partial T}{\partial t} + \rho_l C_{pl} U \cdot \nabla T = \nabla \cdot (k_l \nabla T) \tag{3}$$

$$\rho_s C_{ps} \frac{\partial T}{\partial t} = \nabla \cdot (k_s \nabla T) \tag{4}$$

The phase change process of the paraffin is characterized by the variable θ given in Eq. 5. Note that this variable represents the fraction of paraffin in liquid phase.

$$\theta = \begin{cases} 0, T < T_{m1} \\ \frac{(T - T_{m1})}{(T_{m2} - T_{m1})}, T_{m1} \le T < T_{m2} \\ 1, T_{m2} \le T \end{cases} \tag{5}$$

Mesh Sensibility and Methods

3D model geometry and heat transfer simulation are made with COMSOL Multiphysics Software. The calculation is performed using the finite element method (FEM). Mesh sensitivity is studied by using several tetrahedral elements' number of 34759, 66363, and 122060 with a time step of 0.1 s. The average errors between the first and the second mesh grids and the second and the third mesh grids are 0.66% and 0.07%, respectively. Then 66363 domain elements number is considered enough to ensure the convergence and computing precision. The convergence of the solution is verified at each time step and the iterations were stopped when the maximum difference between two successive iterations is less than 10^{-2}.

Numerical Results

Figure 5 shows a comparison between the numerical results and experimental data for the PCM/MF composite material. The symbol "x" denotes the horizontal coordinate of the numerical model and the heat flux direction. The coordinates x = 0 mm and x = 5 mm refer to T2 and T3 positions, respectively. The obtained experimental temperatures of PCM/MF composite at T2 position are supposed as a boundary condition in x = 0 mm, where numerical and experimental temperature curves are taken equals. For the opposite face of PCM/MF composite (in x = 5 mm), both numerical and experimental results show that PCMs begin to melt around t = 2500 s

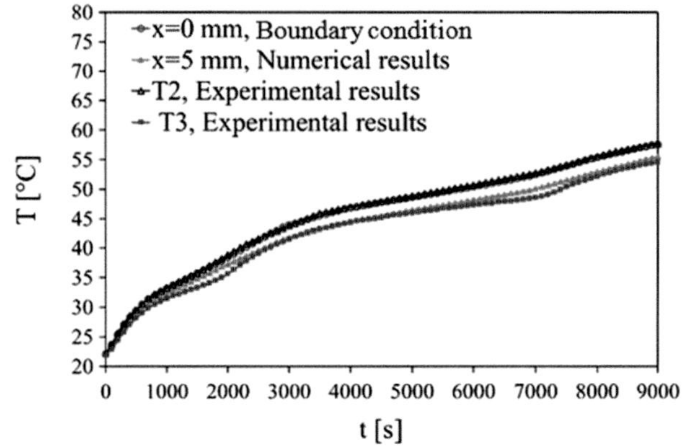

Fig. 5 Numerical versus experimental results for PCM/MF composite

and the phase change is completed approximately at t = 7000 s, and a good agreement between them is achieved. Also, it can be seen that the results are closed to each other with a maximum and an average relative variation of 5.3% and 1.8%, respectively.

To understand the heat transfer through the composite material, the transient temperature field for PCM/MF, with porosity of 96.1% and pore diameter of 2.45 mm, is shown in Fig. 6. By observing the temperature field profile at different moments from 2800 to 3100 s, one can easily note that the heat diffusion is faster in copper

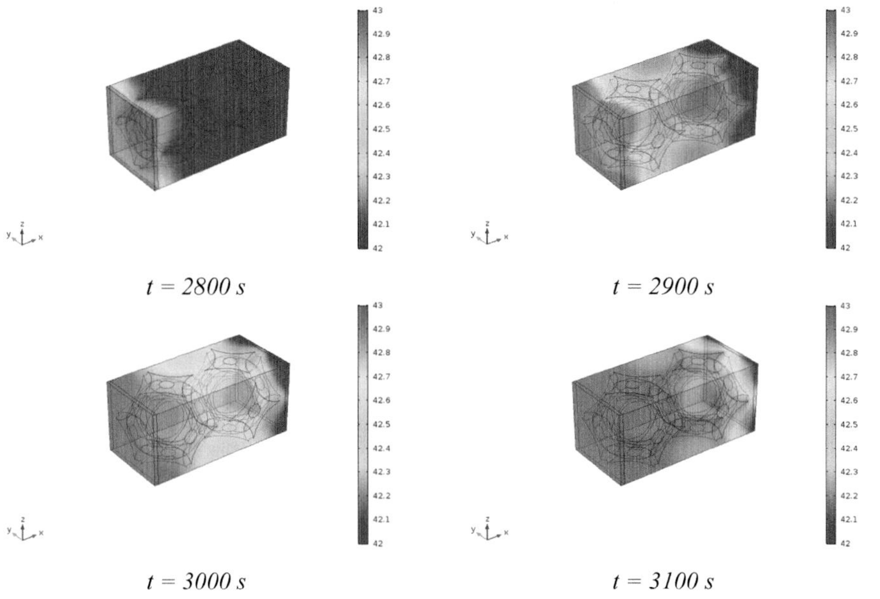

Fig. 6 Paraffin melting front imbedded in copper foam (ε = 98%, dp = 2.5 mm) for different moments

foam than in paraffin, and this is due to the high thermal conductivity of copper. Consequently, the paraffin fusion front is not uniform in this case of composite heat sink especially for the advanced times.

Conclusion

In this study, the heat transfer through a paraffin/copper foam composite which is used as a heat sink is investigated experimentally and numerically for electronic devices. In the experimental setup, the paraffin/copper foam composite is heated to permit paraffin changing state from solid-to-liquid phase. The obtained results show that PCM/MF composite material drops the heat source overheating by about 34% compared to the case without heat sink device. In addition, the composite material improves heat diffusion by about 52% compared to the pure paraffin case. A microcellular model as a body-centered cubic geometry (BCC) is developed using the COMSOL Multiphysics Software. The obtained numerical results are found in good agreement with the experimental data and show that the insertion of the copper foam is good to improve the heat distribution inside the microcellular composite model.

References

1. Zhang P, Meng Z, Zhu H, Wang Y, Peng S (2015) Experimental and numerical study of heat transfer characteristics of a paraffin/metal foam composite PCM. Energy Procedia 75:3091–3097
2. Drissi S, Ling TC, Mo KH (2019) Thermal efficiency and durability performances of paraffinic phase change materials with enhanced thermal conductivity–a review. Thermochim Acta 673:198–210
3. Younsi Z, Joulin A, Zalewski L, Lassue S, Rousse D (2009) Analyse numérique de la fusion de matériaux à changement de phase dans une enceinte rectangulaire chauffée par une paroi latérale. In: Proceedings of Colloque Inter-Universitaire Franco-Québecois. (Lille, France, 18–20 May 2009)
4. Joseph A, Kabbara M, Groulx D, Allred P, White MA (2016) Characterization and real-time testing of phase-change materials for solar thermal energy storage. Int J Energy Res 40–1:61–70
5. Scott RP, Ahmed T, Groulx D (2016) Thermal management of portable electronics using phase change materials: initial experiments. In: Proceeding of rim thermal engineering conference. Hawaii's Big Island, USA, pp 13–17
6. Sponagle B, Groulx D (2015) Thermal modeling of tablets: temperature management using phase change materials. In: Proceeding of thermal fluid engineering summer conference. New York City, USA, pp 9–12
7. Lachheb M, Karkri M, Albouchi F, Mzali F, Nasrallah SB (2014) Thermophysical properties estimation of paraffin/graphite composite phase change material using an inverse method. Energy Convers Manag 82:229–237
8. Alshaer WG, Nada SA, Rady MA, Le Bot C, Palomo Del Barrio E (2015) Numerical investigations of using carbon foam/PCM/Nano carbon tubes composites in thermal management of electronic equipment. Energy Convers Manag 89:873–884

9. Fok SC, Shen W, Tan FL (2010) Cooling of portable hand-held electronic devices using phase change materials in finned heat sinks. Int J Therm Sci 49–1:109–117
10. Zhao CY, Lu W, Tian Y (2010) Heat transfer enhancement for thermal energy storage using metal foams embedded within phase change materials (PCMs). Sol Energy 84–8:1402–1412
11. Xiao X, Zhang P, Li M (2013) Preparation and thermal characterization of paraffin/metal foam composite phase change material. Appl Energ 112:1357–1366
12. Banhart J (2001) Manufacture, characterisation and application of cellular metals and metal foams. Prog Mater Sci 46–6:559–632

Cooling of PEM Fuel Cell Stacks Using Open-Cell Metal Foam

Ali A. Hmad and Nihad Dukhan

Abstract For safe and efficient operation of the proton exchange membrane fuel cell (PEMFC), effective thermal management is needed. This study presents the use of aluminum foam in PEMFC's thermal management using forced air. An analytical study was performed using a bipolar plate air-cooled channel to examine the effects of operating temperatures on fuel cells and to study heat transfer and temperature distribution in the stack. In the simulation, the heat transfer and fluid flow for a thin layer of metal foam inserted between two bipolar plates of fuel cell stack were simulated using ANSYS-Fluent 19.2, and both the local thermal equilibrium and the local thermal non-equilibrium assumptions were applied. The outcomes, such as pressure drop through porous medium per plate length and plate maximum temperatures, were calculated. The temperature profile distributions were drawn, and the comparison between two models was investigated. Based on this simulation, a new design was provided to improve the thermal design of an air-cooled fuel cell by using metal foam.

Keywords Metal foam · Bipolar plate · PEM fuel cells · Air cooling · 2D simulations

Nomenclature

C_p Heat capacity of fluid (J/kg K)
K Thermal conductivity (W/m K)
P_r Prandtl number of air (dimensionless)
ppi Number of pores per inch

A. A. Hmad (✉) · N. Dukhan
Department of Mechanical Engineering, University of Detroit Mercy, 4001 W. McNichols Rd, Detroit, MI 48221, USA
e-mail: Hmadaa@udmercy.edu

N. Dukhan
e-mail: dukhanni@udmercy.edu

N. Dukhan (ed.), *Proceedings of the 11th International Conference on Porous Metals and Metallic Foams (MetFoam 2019)*, The Minerals, Metals & Materials Series, https://doi.org/10.1007/978-3-030-42798-6_16

R_{ep} Reynolds number (dimensionless)
d_p Pore diameter (m)
C Inertial loss coefficient (m^{-1})
Δ_p Pressure drop (pa)
L Sample length in flow direction (m)
h_{sf} Interfacial heat transfer coefficient (W/m^2K)
q_h Heat flux (W/m^2)
T Temperature (K)
V Inlet fluid velocity (m/s)
k_e Effective thermal conductivity (W/m K)
u Average velocity (m/s)
x Axial coordinate along the flow direction (m)
y Transverse coordinate along foam sample (m)

Greek symbols

σ Surface area density
ρ Density of fluid (kg/m^3)
ε Porosity
μ Dynamic viscosity (kg/m.s)

Subscripts

f Fluid
s Solid

Introduction

Fuel cells with a higher power density produce excess heat, which needs to be removed. When the waste heat in Proton Exchange Membrane Fuel Cells (PEMFCs) rises, it dries the membrane out and increases the ohmic resistance [1]. Designable thermal management system needs to be applied; different heat-removing cooling systems are used in PEMFCs [2]. The current cooling systems are designed to have channels machined in the bipolar plates located on both sides of cathode and anode with circulated coolants [3]. Thus, evaporative cooling methods are used to inject water as a liquid directly into the system to keep the membrane cooled and humidified [4]. Lately, heat pipe systems have been considered in PEMFC thermal management [5]. Dukhan and Chen [6] experimentally and analytically studied the heat transfer for different cases of metal foam with different properties and at varied velocities subjected to heat flux from the top. Generally, modeling porous medium in heat transfer is done using two methods. First is the local thermal equilibrium (LTE) assumption, in which the fluid and solid have similar temperatures. This model uses only a single energy equation. Second, the thermal method is the local thermal non-equilibrium model (LTNE) assumption in which the two-model equation is utilized [7]. The following studies demonstrate how these methods are applied.

More specifically, metal foams have been used in PEMFC fuel cells to enhance heat transfer. The heat equation had been solved by Shimpalee and Dutta [8] when they designed a 3D model to study the distribution of temperature inside a PEMFC channel. Another 3D model was designed by Sinha and Wang [9] analyzed heat and current transport in a stack single channel and studied the effect of high operating temperature on the performance of PEMFC. Odabaee et al. [10] investigated the possibility of using metal foams with air-cooled systems in PEM fuel cell system applications instead of using the water-cooled heat exchangers system. Santamaria et al. [11] investigated numerically and experimentally a cooling micro-channel filled with nickel foam for PEMFC. Significantly, the temperature drop in nickel foam micro-channel model was detected. Recently, Vazifeshenas et al. [12] studied the effect of using metal foam with different porosities inserted in the PEM cooling channel on the stack performance. The results proved that utilizing metal foam increased the heat transfer and pressure drop. This study presents a simulation of the air-cooled channel filled with metal foam to examine the enhancement of heat transfer and pressure drop through the porous media and to evaluate the effects of operating temperatures on stacks. The study also investigates the difference between the local thermal equilibrium model (LTE) and the local thermal non-equilibrium model (LTNE) in forced convection flow cross metal foam.

Modeling

Geometry

The schematic view of a 2D simple air channel filled with metal foam was built as shown in Fig. 1. All shoulders between air channels were removed and filled with aluminum foam at both sides of the anode and cathode. Air was forced into the inlet

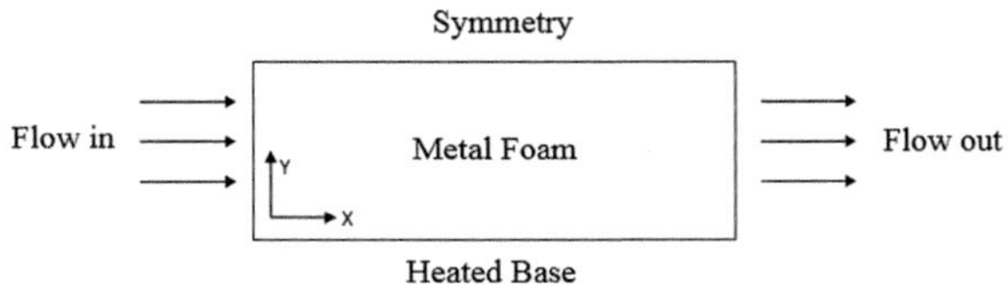

Fig. 1 Schematic view of a simple air channel filled with metal foam

Table 1 Metal foam properties

Material	Pores per inch (ppi)	$L \times W \times H$ (mm)	Porosity ε (%)	Pore diameter dp (mm)
Aluminum	40	50 × 50 × 5	60, 70, 80, 90	0.625

through the metal foam, and constant heat flux was subjected to the heated plate. Metal foam properties are described in Table 1.

Governing Equations

Continuity equation:

For the continuity equation, the fluid flow was supposed to be steady and incompressible through the porous region. The continuity equation is given by Eq. 1.

$$\frac{\partial u}{\partial x} = 0 \tag{1}$$

Momentum equation:

The steady state of momentum equation is represented by Eq. 2.

$$\frac{\partial P}{\partial Y} + \frac{\mu \partial^2 u}{\varepsilon \partial x^2} - \frac{\mu}{k} u = \frac{\rho f}{k} u^2 = 0 \tag{2}$$

where ε and k are the porosity and the permeability of the medium, respectively. It was observed that both third and fourth terms in Eq. 2 represent the pressure drop.

Thermal equations:

LTE model:

The single energy equation is used in this assumption. The effective thermal conductivity of the metal foam k_e affects the results of energy equation in the LTE model. The energy equation is given by Eq. 3.

$$\left(\rho C_p\right)_f u \cdot \frac{\partial T}{\partial Y} = k_e \left(\frac{\partial^2 T}{\partial y^2} + \frac{\partial^2 T}{\partial x^2} \right) \tag{3}$$

LTNE model:

Fluid and solid energy equations for LTNE model are given by Eqs. 4 and 5.

Energy equation for fluid:

$$\varepsilon k_f \left[\frac{\partial^2 T_f}{\partial y^2} + \frac{\partial^2 T_f}{\partial x^2} \right] + h_{sf} \sigma \left(T_s - T_f \right) = 0 \tag{4}$$

Energy equation for solid:

$$(1-\varepsilon)k_f\left[\frac{\partial^2 T_s}{\partial y^2}+\frac{\partial^2 T_s}{\partial x^2}\right]+h_{sf}\sigma\left(T_f-T_s\right)=0 \tag{5}$$

where h_{sf} is the interfacial heat transfer coefficient and σ is the interfacial area density.

Boundary Conditions

Channel inlet:

$$u=u_{\text{in}} \text{ and } T_{\text{f}}=T_{\text{in}} \tag{6}$$

Symmetry wall:

$$\frac{\partial T_f}{\partial y}=\frac{\partial T_S}{\partial y}=0 \tag{7}$$

Heated base:

$$q_h=-k_f\frac{\partial T_f}{\partial y}=-k_s\frac{\partial T_s}{\partial y} \tag{8}$$

Channel outlet:

$$\frac{\partial T_f}{\partial x}=\frac{\partial T_s}{\partial x}=0 \tag{9}$$

Numerical Model

Physical Geometry

To provide an example of air flow through porous media, a simulation was built. A 2D typical geometry was created in ANSYS-Fluent 19.2 to investigate the pressure drop and temperature distribution. A rectangular channel of metal foam filled between two heated plates of aluminum with dimensions of 50 mm × 5 mm was designed. Air was forced to flow in the x-direction through a 40 ppi aluminum foam.

LTNE Modeling

A solid zone that is spatially coincident with the fluid zone had to be created. Both interfacial area density and heat transfer coefficient were calculated and manually entered in CFD to solve the thermal equations. Permeability and inertial loss

Table 2 Permeability, inertial loss coefficient, and surface area density for 40 ppi metal foam

Pores per inch (ppi)	Porosity ε (%)	Permeability K (m^2)	Inertial loss coefficient C (m^{-1})	Surface area density σ (m^2/m^3)
	60	3.51E-09	483.84	2943.56
40	70	9.92E-09	576.24	2743.74
	80	3.33E-08	573.44	2462.12
	90	1.89E-07	408.24	1980.68

coefficient also needed to be calculated according to Eqs. 10 and 11.

$$K = \frac{\varepsilon^3 D_p^2}{150(1-\varepsilon)^2} \tag{10}$$

$$\text{Iner. Resis} = \frac{3.5(1-\varepsilon)}{D_p \varepsilon^3} \tag{11}$$

Dukhan and Chen [6] calculated the interfacial area density for 40 ppi of metal foam. They used a correlation provided by ERG Materials and Aerospace given by Eq. 12.

$$\sigma = 694.57 \ln(1-\varepsilon) + 3579.99 \tag{12}$$

where σ is the surface area per unit volume of the foam. Both interfacial area density and heat transfer coefficient needed to be calculated and manually entered in CFD to solve the thermal equations. Kuwahara et al. [13] numerically studied the coefficient of interfacial convective heat transfer in porous media given by Eq. 13.

$$h_{\text{sf}} = \frac{k_f}{d_p}\left[1 + \frac{4(1-\varepsilon)}{\varepsilon}\right] + \frac{1}{2}(1-\varepsilon)^{1/2}\text{Re}_p^{0.6}\text{Pr}^{1/3} \tag{13}$$

Table 2 shows the permeability, inertial loss coefficient, and surface area density for 40 ppi metal foam.

Results and Discussion

Pressure Drop Per Length

In this study, the permeability and form drag coefficient were important parameters to measure the pressure drop across metal foam. The measurements of pressure drop across the metal foam are calculated at different velocities from 0.2 m/s to 3 m/s and used Forchheimer equation as given by Eq. 14.

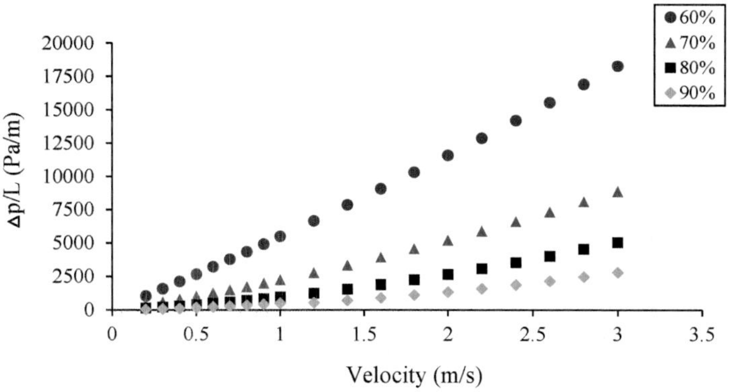

Fig. 2 Pressure drop per length of 40 ppi for different inlet fluid velocities and porosities

$$\frac{\Delta p}{L} = \frac{\mu}{k} V + \rho C V^2 \tag{14}$$

where $\underline{K}$ and C are the permeability and form drag coefficient, respectively. Figure 2 shows the pressure drop differences per length at different inlet fluid velocities. Comparing the pressure drop between the four cases, the pressure drop at 60% porosity is higher compared to only half that amount at the same inlet velocity.

In the second case with porosity of 70%, it can be observed that pressure drop is correlated with metal foam porosity. A high flow resistance is accrued when the channel is filled with 6101-T6 aluminum foam, which shows a higher pressure drop due to its low porosity and low permeability. That means an increase in porosity of aluminum foam results in lower pressure drop.

LTNE Results

The homogeneous 40 ppi aluminum alloy 6101-T6 foam was chosen to be modeled, with porosities of 60, 70, 80, and 90%, and inlet velocities from 0.2 to 3 m/s. The safe range of operating temperatures is from 328 to 363 K. Figure 3 illustrates the temperature distribution for 40 ppi with 60% porosity at inlet velocities 2 and 2.6 m/s increasing the velocity from 2 to 2.6 m/s and decreasing the heated plate temperature to 5 degrees in temperature difference.

Figure 4 shows the maximum plate temperatures for the four cases; the model with porosity of 90% has the largest range of temperature difference between the inlet and outlet channels. The fluid velocity and heat transfer between metal foam and the air dropped the maximum plate temperatures to the safe range. The heat absorption in the flow region increases with the increase of the fluid flow through the metal foam. The effect of porosity was clearly observed, and increasing metal foam

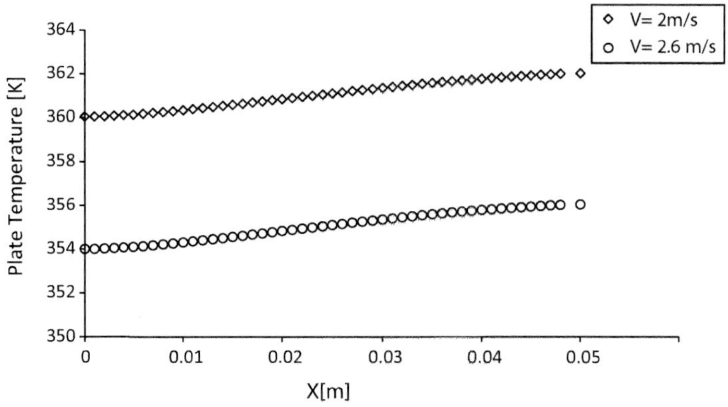

Fig. 3 The temperature profiles for 40 ppi with 60% porosity at inlet velocities 2 and 2.6 m/s

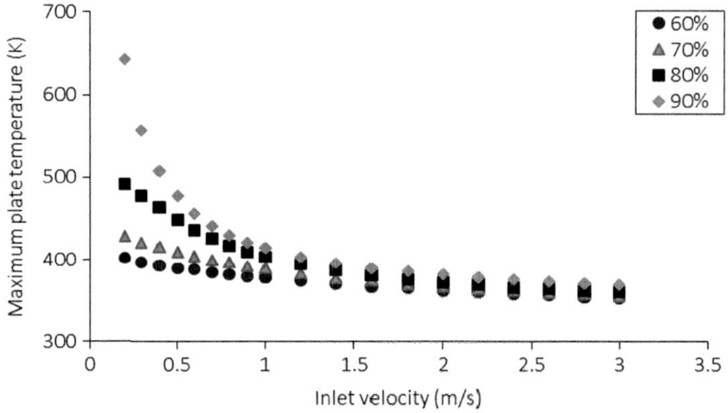

Fig. 4 Maximum plate temperatures at different inlet velocities for 40 ppi with different porosities

porosity increases the temperature difference between inlet and outlet temperatures which resulted in more heat being absorbed into the metal foam region.

Comparison Between LTE and LTNE Models

The results also showed the difference between LTE and LTNE models, when it comes to the plate temperature along the channel length. In the LTE model, both air flow and plate temperatures were almost similar.

Conclusion

The performance of using a bipolar plate air-cooled channel filled with metal foam was analytically studied. A 2D model was built to examine the effects of operating temperature on PEM stacks, and to study heat transfer and pressure drop through the foam. Utilizing 40 ppi aluminum metal foam with different porosities to evaluate the pressure drop and heat transfer. The metal foam thermal properties, such as the permeability and inertial coefficient, were calculated analytically. The model was built to investigate using metal foam in thermal management of PEMFC. The results showed that compared with LTE and LTNE assumptions, the air-cooled channel filled with metal foam significantly maintains the operating temperatures for PEMFC stack within the safe range. The new model can be applied to direct the guidance for future experiments to optimize the thermal management of fuel cell stacks.

References

1. Paserin V, Shu J, Renny, L, Liu AM, Yu BQ, Groo M (2007) Commercial production and applications of Nickel based specialty foams. In: Porous metals and metallic foams: proceedings of the fifth international conference on porous metals and metallic foams, 5–7 Sept 2007, Montreal, Canada
2. Kandlikar S, Lu Z (2009) Thermal management issues in a PEMFC stack—a brief review of current status. Appl Therm Eng 29:1276–1280
3. Brambilla M, Mazzucchelli G (2004) Fuel cell with cooling system based on direct injection of liquid water. US Patent 6,835,477, assigned to Nuvera Fuel Cells Europe S.r.l
4. Wang J, Shi M (2006) Study on two-phase countercurrent flow and transport phenomenon in PEM of a direct methanol fuel cell. Sci China E: Technol Sci 49:102–114
5. Clement J, Wang X (2013) Experimental investigation of pulsating heat pipe performance with regard to fuel cell cooling application. Appl Thermal Energy 50:268–274
6. Dukhan N, Chen KC (2007) Heat transfer measurements in metal foam subjected to constant heat flux. Exp Therm Fluid Sci 32(2):624–631
7. Gangapatnam P, Kurian R, Venkateshan SP (2018) Numerical simulation of heat transfer in metal foams. Heat Mass Transf/Waerme- und Stoffuebertragung 54(2):553–562
8. Shimpalee S, Dutta S (2000) Numerical prediction of temperature distribution in PEM fuel cells. Numer Heat Transfer Part A 38:111–128
9. Sinha PK, Wang C-Y, Beuscher U (2007) Transport phenomena in elevated temperature PEM fuel cells. J Electrochem Soc 154(1):B106
10. Odabaee M, Hooman K (2013) Metal foam heat exchangers for thermal management of fuel cell systems—an experimental study. Exp Thermal Fluid Sci 51:214–219
11. Santamaria A (2017) Metal foam microchannel heat exchangers for cooling of fuel cells & flow batteries, pp 1–7
12. Vazifeshenas Y, Sedighi K, Shakeri M (2019) Heat transfer in PEM cooling flow field with high porosity metal foam insert. Appl Therm Eng 147(October 2018):81–89
13. Kuwahara F, Shirota M, Nakayama A (2001) A numerical study of interfacial convective heat transfer coefficient in two-energy equation model for convection in porous media. Int J Heat Mass Transf 44(6):1153–1159

Cylinder-Pack Modeling of Open-Cell Metal Foam for Flow and Heat Transfer

Nihad Dukhan and Omer Saad

Abstract Because of its morphology, accessible huge surface, and high porosity, open-cell metal foam has been investigated for flow and heat transfer applications. Capturing the actual knotted internal structure of the foam is rather expensive and time-consuming. Some researchers have relied on geometrical idealization of the actual structure in order to save on computational time and cost. However, some of these idealizations are ad hoc and often do not necessarily exhibit intimate relationship to the actual foam, especially in terms of key properties that control flow and heat transfer in the foam. This study investigates certain staggered arrangement of thin vertical cylinders that can provide the same transport effects as a given foam with known morphology. The average cell diameter, surface area density, and porosity are used to describe the foam. The transverse pitch of the cylinders is chosen such that the surface area density of the foam and that of the cylinders is equal. The diagonal pitch is determined based on the staggering arrangement of the cylinders, while the diameter of a typical cylinder is calculated from the porosity of the foam. By founding a robust, yet minimal, technique for representing the complex morphology of the foam, substantial savings in computational time and effort will be realized.

Keywords Metal foam · Geometric modeling · Flow · Heat transfer

Introduction

Metal foam's production methods, properties, and applications have been covered in [1]. The open-cell type foams have high porosity, permeability, thermal conductivity, and surface area density. The web-like internal structure of metal foams promotes mixing through flowing fluids. As such, metal foams are attractive for heat transfer

---

N. Dukhan (✉) · O. Saad
Department of Mechanical Engineering, University of Detroit Mercy, 4001 W. McNichols Rd., Detroit, MI 48221, USA
e-mail: dukhanni@udmercy.edu

O. Saad
e-mail: omarsasd2003@gmail.com

N. Dukhan (ed.), *Proceedings of the 11th International Conference on Porous Metals and Metallic Foams (MetFoam 2019)*, The Minerals, Metals & Materials Series, https://doi.org/10.1007/978-3-030-42798-6_17

enhancement systems. Flow and heat transfer in metal foam have been investigated by many researchers. One of the experimental studies regarding flow and pressure drop is [2]; as for experimental heat transfer in metal foam, [3–5] should be consulted. Flow and heat transfer in metal foam have also been investigated numerically by solving the governing of momentum and energy equations. One approach is to solve these equations at the microscopic level over the actual internal structure of the foam (captured by microcomputed tomography), see for example [6, 7]. The other approach is to solve a macroscopic form of the governing equations obtained by the so-called volume averaging technique, e.g., [8–10]. Both of these methods are expensive in terms of cost and computational time.

A third approach is to solve the microscopic equations over a simplified geometrical model of the internal structure of the foam. Current geometrical modeling of metal foam involves complex geometries such as the tetrakaidecahedra (a 14-sided figure having 6 quadrilateral and 8 hexagonal faces). Other "simpler" geometric cells can be found in [11–13]. Typically, geometrical modeling studies consider one or only few cells in the flow direction, which is not sufficient for capturing entry and boundary effects. The solutions over these models are also time-consuming and computationally expensive.

In this paper, a simple geometrical modeling techniques is described, which eliminates some of the issues described above. Because of ease-of-use aspect of the proposed technique, studying flow and heat transfer will indeed become more accessible to a wider range of engineers and scientist, which will proliferate the use of such materials. As a matter of fact, the new technique, once validates, can be used to custom design new metal foams with tailored properties for targeted applications. The new modeling can easily accommodate larger domains of metal foam, which leads to more realistic results.

The Cylinder-Pack Model of Open-Cell Metal Foam

When open-cell metal foam is attached to a heated surface, the thin ligaments of the foam provide obstructions to any low. They also conduct heat from the heated boundary, which is subsequently transferred to the moving fluid by convection. The ligaments of metal foam have small diameters compared to cell diameter. For example, for a 97% porous aluminum foam having 10 pores per inch (ppi), the ligament diameter is 410 μm, while the cell diameter is 6900 μm [11]. Therefore, the actual structure of the foam, Fig. 1, can be represented by thin solid cylinders attached to the solid surface. This representation is kept intentionally simple in order to be accessible to a large audience and be easily applied. Thermophysical properties of the cylinders must be the same as those of the foam. The heights of the cylinders must be greater than the thickness of the momentum and thermal boundary layers.

Given a commercial open-cell metal foam having a porosity $\varepsilon_{\mathrm{FOM}}$ and a surface area density σ_{FOM}, an equivalent cylinder pack can be found. A choice (to be verified) of how the cylinders should be arranged can also be made, Fig. 2. For the choice of

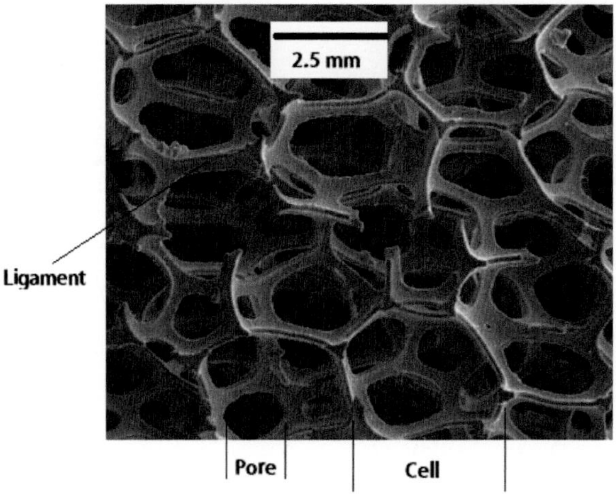

Fig. 1 Actual internal structure of open-cell metal foam

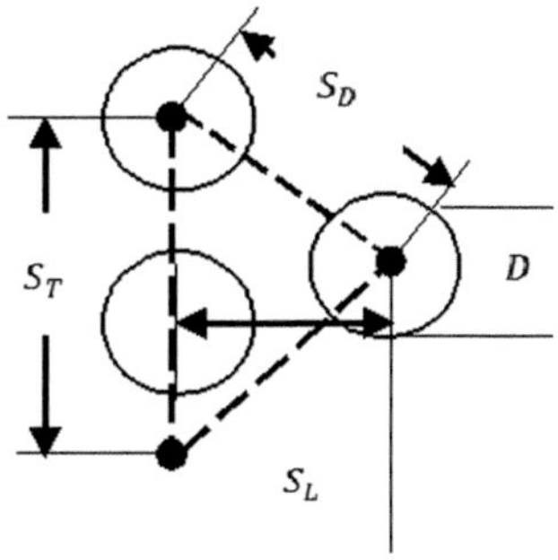

Fig. 2 Equilateral triangle arrangement

equilateral triangle staggered arrangement of thin vertical cylinders, the equivalent system is fully described by the diameter of a cylinder D and diagonal S_D, longitudinal S_L and transverse S_T pitches.

Figure 3 shows a top view of an equivalent cylinder pack bounded at the bottom by a solid wall and having width W, length L, and height H.

Through geometrical arguments and equivalency conditions imposed on the system, the diameter of a cylinder is obtained as

$$D = \frac{4(1 - \varepsilon_{\mathrm{FOM}})}{\sigma_{\mathrm{FOM}}} \tag{1}$$

and the diagonal pitch is given by

$$S_D = \frac{2}{\sigma_{\mathrm{FOM}}}\sqrt{\frac{2\pi(1 - \varepsilon_{\mathrm{FOM}})}{\sqrt{3}}} \tag{2}$$

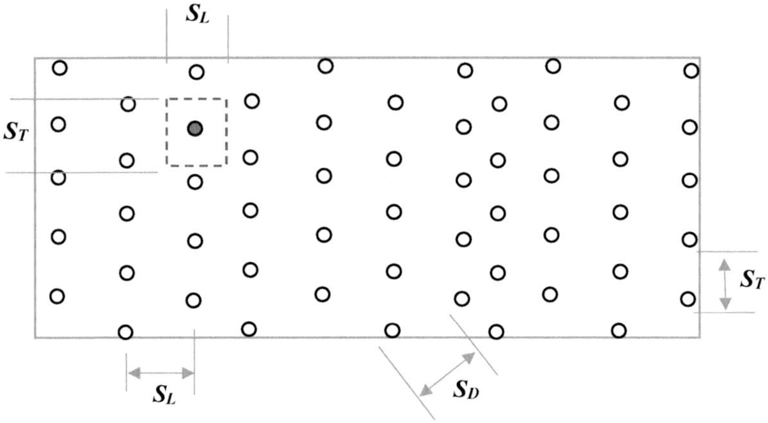

Fig. 3 Top view of a surface covered with a cylinder-pack model

For equilateral triangle arrangement

$$S_D = S_T = (2/\sqrt{3})S_L \tag{3}$$

From Eq. 3, the remaining two pitches are

$$S_L = \frac{1}{\sigma_{\mathrm{FOM}}}\sqrt{2\sqrt{3}\pi(1-\varepsilon_{\mathrm{FOM}})} \tag{4}$$

$$S_T = \frac{2}{\sigma_{\mathrm{FOM}}}\sqrt{\frac{2\pi(1-\varepsilon_{\mathrm{FOM}})}{\sqrt{3}}} \tag{5}$$

Note that in the last four equations, D, S_D, S_L, and S_T (the complete description of the cylinder pack) are expressed in terms of the foam's porosity $\varepsilon_{\mathrm{FOM}}$ and surface area density σ_{FOM} only.

Example

As an example of how a commercial open-cell aluminum foam is represented by a set of cylinders, consider a block of open-cell aluminum foam having 20 pores per inch (ppi) brazed to a solid aluminum base. Let the dimensions of the cross section of the foam be $W = 10.16$ cm (4 in.) and $H = 10.16$ cm (4 in.). The length in the flow direction is $L = 5.08$ cm (2 in.). The block is brazed to a 3-mm-thick solid aluminum base. The porosity is 78.2%, which can be determined by comparing the weight of foam to the weight of an equivalent volume of solid aluminum alloy from which the foam is made. The surface area density is available directly from manufacturers, e.g., for open-cell aluminum foam having 20 ppi [14]:

$$\sigma_{\mathrm{FOM}} = 442.20\ln(1-\varepsilon_{\mathrm{FOM}}) + 2378.62\left(m^2/m^3\right) \tag{6}$$

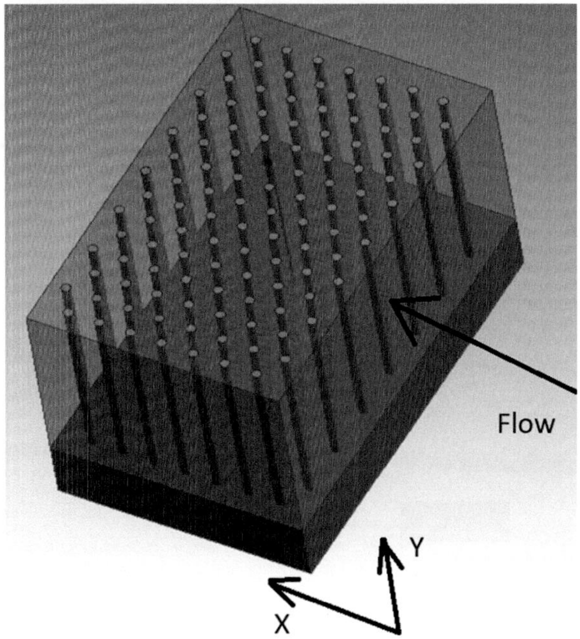

Fig. 4 An equivalent cylinder-pack model for a heat sink metal foam heated from the bottom

The equivalent cylinder pack representing this foam block is obtained using Eq. 6 and represented in Fig. 4. The diameter of each cylinder is 0.41 mm, and diagonal, S_D longitudinal S_L, and transverse S_T pitches are 1.371, 1.187, and 1.372 mm, respectively.

The aluminum block is acting as a heat sink (heated from the bottom with a constant heat flux of 29900 W/m^2) and is cooled by air. The cross-sectional area is open to the flow, while all other sides are insulated. The cylinder pack can be imported into a package, i.e., ANSYS, and the heat transfer problem can be solved. To save time, and due to symmetry, only a thin representative unit having width of 1.372 mm and a height of 50 mm and the full length in the flow direction 50.8 mm was solved. This segment includes 21 cylinders and is shown after meshing in Fig. 5. The mesh for the geometry has 100204 nodes and 117717 elements.

Numerical Results

For an inlet air temperature of 300 K and a velocity of 2.5 m/s, Fig. 6 shows the solid cylinders' temperature as a function of Y (distance from the heated base) at three different axial locations form the inlet X = 6.53 mm, 19.05 mm, and 31.75 mm. The same information is given for a velocity of 2.71 m/s in Fig. 7. It is clear that the temperature behavior at the qualitative level makes sense. The temperature decreases as the distance from the heated base increases. Also, the temperature increases in the

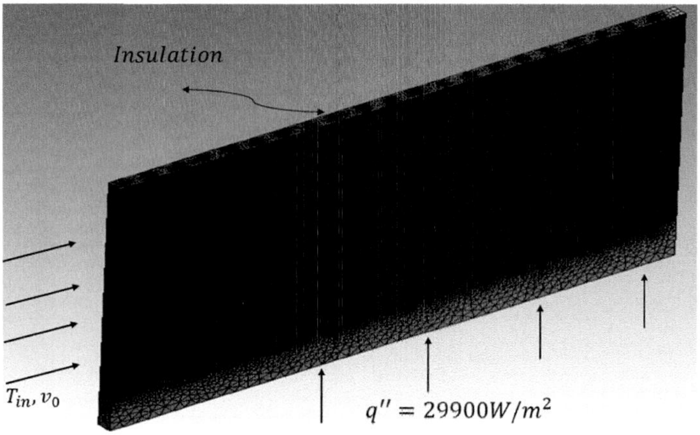

Fig. 5 A representative section as a solution domain with meshing

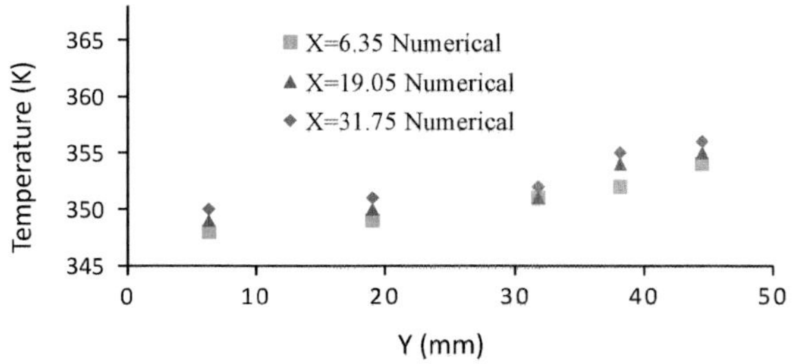

Fig. 6 Temperature distribution for the tube bank at different axial locations for velocity 2.5 m/s

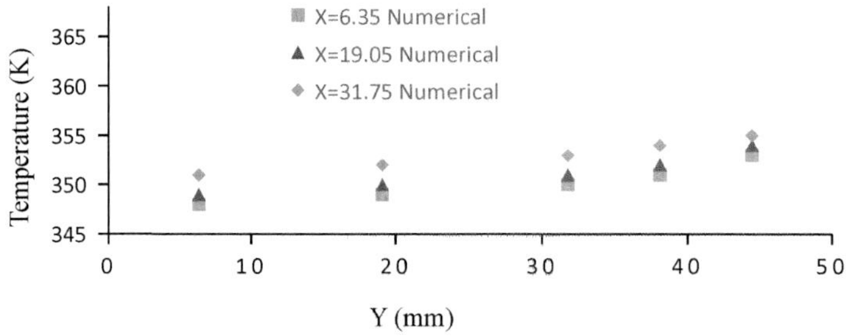

Fig. 7 Temperature distribution for the tube bank at different axial locations for velocity 2.71 m/s

flow direction due to continuous heating of the cooling air as it travels through the heat sink. Nonetheless, direct comparison to heat transfer data in metal foam is needed in order to establish the validity of the modeling approach. Such comparison may lead to ways of improving the model and/or calibrating it. Issues of the permeability and tortuosity may need to be addressed for the foam and the cylinder pack. The effective thermal conductivity and the convective heat transfer coefficient for the foam and the cylinders also need to be considered.

Conclusion

The internal structure of open-cell metal foam has been represented by a cylinder pack or pin-finned surface in equilateral triangle arrangement. The modeling method retains key features of the foam, e.g., volumetric porosity and surface area density. The numerical solution of the heat transfer problem in the fins showed reasonable behavior in terms of dependence on distance along the flow direction and from the heated base. The modeling technique needs to be validated by direct comparison to experimental heat transfer data in metal foam. Also, the pressure drop needs to be investigated and compared to actual experimental data for flow through metal foam. As such, the model may require adjustment and or calibration in order to produce matching results for actual metal foam.

References

1. Dukhan N (ed) (2013) Metal foam: fundamentals and applications. DESTech, Lancaster, PA
2. Dukhan N, Minjeur CA II (2011) A two-permeability approach for assessing flow properties in cellular metals. J. Porous Mat 18(4):417–424
3. Mancin S, Zilion C, Diani A, Rossetto L (2013) Air forced convection through metal foams: experimental results and modeling. Int J Heat Mass Transfer 62:112–123
4. Calmidi VV, Mahajan RL (2000) Forced convection in high porosity metal foams. J Heat Transfer 122:557–565
5. Kurbas I, Celik N (2009) Experimental investigation on forced and mixed convection heat transfer in a foam-filled horizontal rectangular channel. Int J Heat Mass Transfer 52:1313–1325
6. Bodla KK, Murthy JY, Garimella SV (2010) Microtomography-based simulation of transport through open-cell metal foams. Numer Heat Transfer Part A 58:527–544
7. Andrew M, Menke H, Blunt MJ, Bijeljic B (2015) The imaging of dynamic multiphase flow using synchrotron-based X-ray microtomography at reservoir conditions. Transp Porous Media 110:1–24
8. Dukhan N (2012) Analysis of Brinkman-extended Darcy flow in porous media and experimental verification using metal foam. J. Fluids Eng 134(7). https://doi.org/10.1115/1.4005678
9. Xu HJ, Qu ZG, Tao WQ (2011) Analytical solution of forced convective heat transfer in tubes partially filled with metallic foam using the two-equation model. Int J Heat Mass Transf 54:3846–3855
10. Dukhan N, Al-Rammahi MA, Suleiman AS (2013) Fluid temperature measurements inside metal foam and comparison to brinkman-darcy flow convection analysis. Int J Heat Mass 67:877–884
11. Kopanidis A, Theodorakakos A, Gavaises E, Bouris D (2010) 3D numerical simulation of flow and conjugate heat transfer through a pore scale model of high porosity open cell metal foam. Int J Heat Mass Transf 53(12):2539–2550
12. Krishnan S, Murthy JY, Garimella SV (2006) Direct simulation of transport in open-cell metal foam. J Heat Transfer 128(8):793–799

13. Karimian SAM, Straatman AG (2008) CFD study of the hydraulic and thermal behavior of spherical-void-phase porous materials. Int J Heat Fluid Flow 29(1):292–305
14. Dukhan N, Patel P (2008) Equivalent particle diameter and length scale for pressure drop in porous metals. Exp Thermal Fluid Sci 32:1059–1067

Influence of Metal Foam Insert Within a Methanol Steam Reformer

Abou Horeira Abaidi and Brahim Madani

Abstract The aim of this work is to study the influence of the insertion of metallic foam in a methanol steam reforming reactor for the production of hydrogen. This work is focused on two configurations: a reactor with and without metal foam. This is a numerical study carried out under Fluent-Ansys software. Temperature and species profiles within the reactor are given. The results show that the metal foam insert improves the efficiency of the reactor by 16%.

Keywords Methanol steam reformer · Metallic foam · Intensification of heat and mass transfers

Nomenclature

n_1, n_2, n_3	Exponential indexes (–)
C	Specific heat, (J kg^{-1} K^{-1})
C_f	Inertial coefficient (–)
D	Mass diffusion (mol m^{-2} s^{-1})
E	Activation energy (J mol^{-1})
h	Height (m)
K	Permeability (m^2)
k_0	Pre-exponential factor ($\text{mol}_{CH_3OH}\text{kg}_{cat}^{-1}$ s^{-1} kPa$^{-\Sigma ni}$)
L	Length (m)
Le	Lewis number (–)

A. H. Abaidi
Laboratory of Polyphasic Transport and Porous Media (LTPMP), USTHB, BP32 El Alia, 16111 Algiers, Algeria
e-mail: aabaidi@usthb.dz

B. Madani (✉)
Laboratory of Multiphase Transport and Porous Media (LTPMP), Faculty of Mechanical and Process Engineering (FGMGP), University of Sciences and Technology Houari Boumediene (USTHB), BP32 El Alia, 16111 Algiers, Algeria
e-mail: bmadani@usthb.dz

N. Dukhan (ed.), *Proceedings of the 11th International Conference on Porous Metals and Metallic Foams (MetFoam 2019)*, The Minerals, Metals & Materials Series, https://doi.org/10.1007/978-3-030-42798-6_18

P	Pressure (Pa)
P_i	Partial pressure (Pa)
r	Reaction rate (mol m^{-2} s^{-1})
R	Universal gas constant (J mol^{-1} K^{-1})
S	Source term (m^3)
S_i	Rate of chemical appearance and disappearance (mol m^{-2} s^{-1})
T	Temperature (K)
u	Axial velocity (m s^{-1})
v	Transverse velocity (m s^{-1})
w	Mass fraction (–)

Greek Symbols

μ	Dynamic viscosity (kg m^{-1} s^{-1})
ρ	Fluid density (kg m^{-3})
ε	Porosity (–)
ν''	Stoichiometric coefficient of product species (–)
ν'	Stoichiometric coefficient of reactant species (–)
λ	Thermal conductivity (W m^{-1} s^{-1})

Subscripts

e	Effective
f	Fluid
i, j	Reactant and product components
s	Solid

Introduction

Metal foam is an ideal prerequisite for the construction of heat exchangers. Their porosity up to 95% facilitates the passage of gases and liquids. So that the large pore exchange surface of the metal foam with the good conductivity of the metal allows the transmission of a large amount of heat and a weak resistance to the flow. Many papers used the metal foam for intensification of heat and mass transfers within multifunction exchangers. Authors in [1] found that the macro-patterned design is a promising strategy for significantly improving the temperature distribution in a methane steam reforming reactor. The work in [2] consists of cascading metal foams which were used as catalyst supports for constructing a new type of cylindrical laminated methanol steam reforming microreactor for hydrogen production. The obtained results show that when the PPI of the metal foam was increased from 50 to 100, both the methanol conversion and the H_2 production increased gradually. The study in [3] concerns the influence of the cell density of a series of metal monoliths on the performances of a methanol steam reformer "MSR". Here, the metal monoliths are used as supports of Pd/ZnO catalyst with variable quantities. The results indicate that the conversion increases with increasing size of the channels that are accompanied by a decrease in the monoliths cell density, and in addition they have observed that the thermal conductivity increases with increasing cell density and the average temperature increases with decreasing thermal conductivity. The results in [4] indicate that even though there is no significant improvement in methanol conversion with increasing catalyst layer thickness, a greater catalyst layer thickness provides the advantage of reducing high temperature elevations across the reformer length. The methanol steam reforming intensification to enhance the hydrogen production in a multi-channel block type micro-reformer was studied in [5]. The effects of the operating parameters such as reforming temperature, space velocity, and catalyst layer thickness on reforming performance have been investigated. For an optimized design and operating conditions, a reformer unit with a volume of 8 cm^3 produced 170 LH_2/h containing on dry basis 75.0% H_2, 23.5% CO_2, 0.06% CH_3OH, and 1.44% CO at 648 K allowing the production of 218–255 W by using a commercial PEMFC with 80% hydrogen utilization. The results in [6] show that the quantities of the converted methanol and the produced hydrogen increase when the inlet steam to carbon ratio (S / C) is increased. These quantities reach their maximum values for S / C = 1.4. Higher heat supply and the use of more narrow channels all increase the methanol conversion and hydrogen molar flow rate. In addition, Settar et al. [7] have presented a numerical study on the improvement of the performances of a methane steam reforming reactor. The obtained results show that the combination between the catalyst arrangement and the metal foam insertion improves the methanol steam reformer efficiency by approximately 44.6%.

The aim of this work is to study the effects of metallic foam insert on the performances of a methanol steam reformer which is intended to produce hydrogen, from heat and mass transfer and efficiency points of view. The simulation will be conducted using Fluent-Ansys software.

Description of Physical Model

The reactor is 2D model, and it is represented in Fig. 1. It is 10 mm of length and 0.6 mm of height. The direction of flow is along the x-axis and the upper wall where the catalyst is located is defined as an active wall. The lower wall is taken adiabatic. For those simulations, we suppose the temperatures at inlet and the upper walls are 120 °C and 180 °C, respectively. The velocity at the inlet is equal to 0.1 m/s, and the operation pressure is the atmospheric pressure.

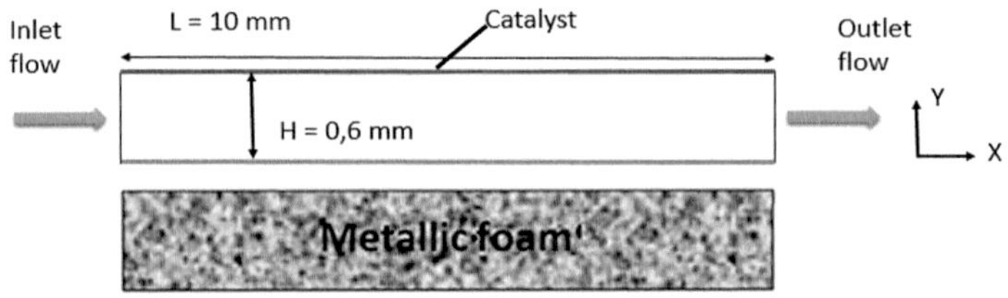

Fig. 1 Diagram of the field of our study in the MSR reactor

In this study, the following assumptions are considered:

- The flow regime is laminar and *steady*.
- Gases are considered ideal gases.
- The pressure drop is zero, and the operating pressure considered is the atmospheric pressure.
- The physical characteristics of the gases are constant.
- The gases can be modeled as incompressible fluids (of constant density). In microscopic and macroscopic calculations, this hypothesis is considered for pressure changes less than 20% of the operating pressure [8].

Kinetics Laws of the Methanol Steam Reforming

Methanol Steam Reforming (MSR) is best carried out over copper-based catalysts as $Cu/ZnO/Al_2O_3$. The global reaction of MSR is given as follows:

$$CH_3OH + H_2O \rightarrow CO_2 + 3H_2 \quad \Delta H_{MSR} = 49.5\,\text{KJ/mo1} \quad (\text{MSR}) \tag{1}$$

In the literature, this reaction is achieved via several mechanisms. In [9] and [10], it is assumed that carbon monoxide is first generated through reaction called methanol decomposition (DM) and then it reacts with water to form carbon dioxide via reaction named water gas shift reaction (WGS):

$$CH_3OH \rightarrow CO + 2H_2 \quad \Delta H_{DM} = 90.5\,\text{KJ/mol} \quad (\text{MD}) \tag{2}$$

$$CO + H_2O \leftrightarrow CO_2 + H_2 \quad \Delta H_{WGS} = -41\,\text{KJ/mol} \quad (\text{WGS}) \tag{3}$$

Jian et al. [11], in their work, note that the measured amount of the by-product CO during the MSR reaction is insignificant at moderate methanol conversion. Thus, upon their experiences they proposed a mechanism as follows:

$$2CH_3OH \rightarrow CH_3OCHO + 2H_2 \quad \text{Methanol dehydrogenation} \tag{4}$$

$$CH_3OCHO + H_2O \rightarrow HCOOH + CH_3OH \quad \text{Methyl formate hydrolysis} \tag{5}$$

$$HCOOH \rightarrow CO_2 + H_2 \quad \text{Formic acid decomposition} \tag{8}$$

Recent studies as that of [12] and [13] confirmed that CO is produced in MSR process via the reverse WGS reaction.

$$CO_2 + H_2 \leftrightarrow CO + H_2O \quad \text{Reverse WGS} \tag{9}$$

In the present work, the rate of the reaction is given by Eq. 10 which is established experimentally in [11]:

$$r_{\text{msr}} = k_0 \exp\left(\frac{-E}{RT}\right) P_{\text{CH}_3\text{OH}}^{n_1} P_{\text{H}_2\text{O}}^{n_2} P_{\text{H}_2}^{n_3} | \tag{10}$$

where $k_0 = 5.31 \times 10^{12}$ $\text{mol}_{\text{CH}_3\text{OH}}\text{kg}_{\text{cat}}^{-1}\,\text{s}^{-1}\,\text{kPa}^{-\Sigma\text{ni}}$, $E = 105$ kJ/mol, $n_1 = 0.26$, $n_2 = 0.03$, $n_3 = -0.2$.

Mathematical Model and Numerical Procedure

The Macroscopic Transport Equations

The fluid flow field is governed by the Navier–Stokes equations and for the temperature and mass fields are, respectively, governed by the energy equation and the equation of species. For the porous region (with an inserted porous media), the macroscopic governing equations, Eqs. (11)–(15), are obtained for representative elementary volume <VER>. These equations are detailed in [14–16]. The phenomenological equations for the smooth channel could be obtained by assuming $Kp \rightarrow \infty$, $\varepsilon \rightarrow 1$ and the effective properties by those of the fluid.

(a) The continuity equation:

$$\frac{\partial \rho_e u}{\partial x} + \frac{\partial \rho_e v}{\partial y} = 0 \tag{11}$$

(b) The momentum equations:

- Following the X-direction:

$$\frac{1}{\varepsilon^2}\left(u\frac{\partial \rho_e u}{\partial x} + v\frac{\partial \rho_e u}{\partial y}\right) = -\frac{\partial p}{\partial x} + \frac{1}{\varepsilon}\left(\frac{\partial}{\partial x}\left(\mu\frac{\partial u}{\partial x}\right) + \frac{\partial}{\partial x}\left(\mu\frac{\partial u}{\partial y}\right)\right) - \left(\frac{\mu}{K_p} + \frac{\rho_e C_f}{\sqrt{K_p}}|u|\right)u \tag{12}$$

- Following the Y-direction:

$$\frac{1}{\varepsilon^2}\left(u\frac{\partial\rho_e v}{\partial x}+v\frac{\partial\rho_e v}{\partial y}\right)=-\frac{\partial p}{\partial y}+\frac{1}{\varepsilon}\left(\frac{\partial}{\partial x}\left(\mu\frac{\partial v}{\partial x}\right)+\frac{\partial}{\partial x}\left(\mu\frac{\partial v}{\partial y}\right)\right)-\left(\frac{\mu}{K_p}+\frac{\rho_e C_f}{\sqrt{K_p}}|u|\right)v| \tag{13}$$

where K_p is the permeability of metal foam and C_f is the Forchheimer coefficient.

(c) The energy equation:

$$\left(u\frac{\partial(\rho C_p)_e T}{\partial x}+v\frac{\partial(\rho C_p)_e T}{\partial y}\right)=\frac{\partial}{\partial x}\left(\lambda_e\frac{\partial T}{\partial x}\right)+\frac{\partial}{\partial x}\left(\lambda_e\frac{\partial T}{\partial y}\right)+S_{\mathrm{WGS}} \tag{14}$$

(d) The species equation:

$$u\frac{\partial\rho_e w_i}{\partial x}+v\frac{\partial\rho_e w_i}{\partial y}=\frac{\partial}{\partial x}\left(D_{i,e}\frac{\partial\rho_e w_i}{\partial x}\right)+\frac{\partial}{\partial x}\left(D_{i,e}\frac{\partial\rho_e w_i}{\partial y}\right) \tag{15}$$

where ρ_e and C_{pe} are the effective density and effective specific heat for the fluid–solid pair.

$$\rho_e=\varepsilon\cdot\rho_f+(1-\varepsilon)\rho_s \tag{16}$$

$$C_{p_e}=\varepsilon\cdot C_{p_f}+(1-\varepsilon)C_{p_S} \tag{17}$$

where $\boldsymbol{D}_i$ represents the mass diffusivity of the species ($i = CH_3OH$, CO_2, H_2O, H_2, CO_2).

The mass-diffusion model in this work is based on the Lewis number which is considered equal to the unit for all the species.

The gas mixture is assumed to be ideal gas; therefore, the species are adapted to the Dalton law (*partial pressure law for perfect gas mixtures*). The partial pressures of the components can be expressed in terms of their molar fraction in the mixture.

The mass diffusion in the porous medium (Eq. 8) and the other mixture properties are given in [17]:

$$D_{i,e}=\frac{\lambda_e}{Le_i(\rho\cdot C_p)_e} \tag{18}$$

The effective thermal conductivity is given in [18, 19]:

$$\lambda_e=0.35(\varepsilon\lambda_f+(1-\varepsilon)\lambda_s)+\frac{0.65}{\left(\frac{\varepsilon}{\lambda_f}+\frac{(1-\varepsilon)}{\lambda_s}\right)} \tag{19}$$

Table 1 Physical characteristics of the porous matrix of our work according to [20]

ε	PPI	d_f (m)	d_p (m)	C_f (–)	K_p (m^2)	ρ_s (kg/m^3)	λ_s (W m^{-1} K^{-1})	Cp_s (J kg^{-1} K^{-1})
0.8991	10	0.00043	0.0032	0.0068	0.94×10^{-7}	850	400	385

Table 2 The boundary conditions of our model

	U	V	T	w
Inlet	1.5×10^{-4} m/s	0	120 °C (393 K)	$w_{CH_3OH} = 0.62$ $w_{H_2O} = 0.38$
Outlet	$\frac{dU}{dx} = 0$	0	$\frac{dT}{dx} = 0$	$\frac{dw_i}{dx} = 0$
Lower surface	0	0	$\frac{dT}{dy} = 0$	$\frac{dw_i}{dy} = 0$
Upper surface	0	0	180 °C (453 K)	$\frac{dw_i}{dx} = -\frac{S_i M_i}{\rho D_{i,eff}}$ $S_i = r(\gamma'' - \gamma')$

The physical characteristics of the matrix layer are summarized in Table 1.

Boundary Conditions

The boundary conditions to complete the mathematical model are given below:

- The walls are considered impermeable.
- The catalytic layer is an active wall and the adaptation of a surface-type reaction approach.
- The effect of the thickness of catalyst is supposed negligible.
- The boundary conditions for simulations are given in Table 2.

Numerical Procedure

The simulation is done under Fluent Fluid Dynamics software which is based on the finite volume method. The SIMPLE algorithm (pressure–speed coupling) was used to solve convection–diffusion equations. A structured meshing composed of square elements is used to mesh the channel.

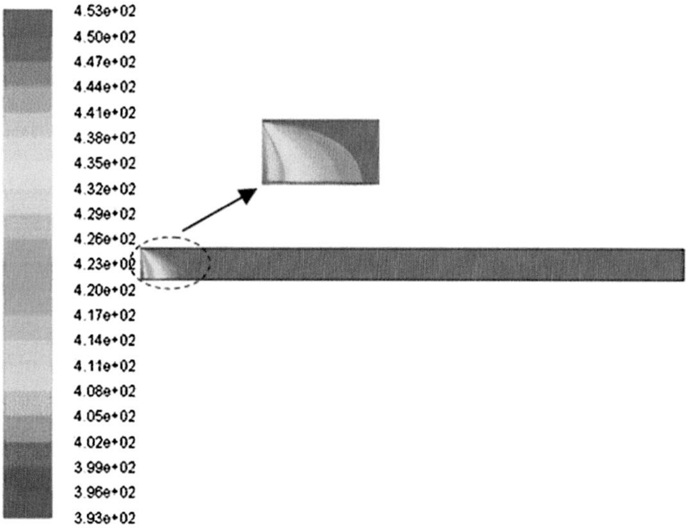

Fig. 2 Temperature distributions along the reactor channel

Results and Discussion

Smooth Channel

Temperature Profile

Figure 2 shows the temperature distribution within the methanol steam reforming reactor. The figure shows the temperature assumed to be 120 °C (393 K) at the inlet of the channel and a temperature which is maintained constant throughout the remainder of the channel at 453 K after initiation of the catalytic reaction. This is due essentially to the continuous and homogeneous supply applied to the hot wall.

Distribution of Chemical Species

Figure 3 shows the molar fraction distributions of chemical species CH_3OH, H_2O, H_2, and CO_2, respectively, in the reactor during the methanol steam reforming reaction as a function of the length of the lower wall of the reactor. These curves show a decrease in the reactants CH_3OH and H_2O accompanied by an increase in the products which are H_2 and CO_2.

However, it is found that the molar production of hydrogen is equal to three times the molar production of carbon dioxide in the same direction as the stoichiometric coefficients assigned to the products H_2 and CO_2 which are, respectively, 3 and 1. Note that the mole fraction's rapid evolution of reactants and products is favored by

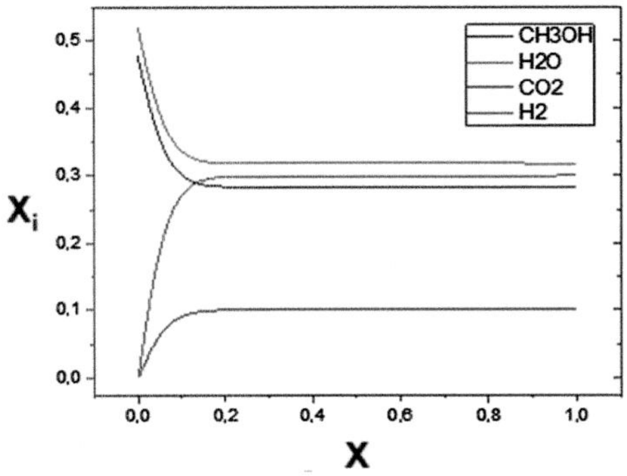

Fig. 3 Molar fraction distributions along the reactor channel

the low value of the axial velocity which is of the order of 10^{-4} ms^{-1}. These molar fractions become constant along the remaining length of the reactor. In this part of the channel, the flow regime is established and the consumption of the reactants becomes maximum.

Reactor Performance with Matrix Layer

According to this figure, the efficiency of this reactor is calculated according to the formula of [21]:

$$\eta = \frac{M_{CH_3OH,in} - M_{CH_3OH,out}}{M_{CH_3OH,in}} = 42\% \tag{20}$$

Channel with Metal Foam

Temperature Profile

Figure 4 shows the structure of the temperature within the reactor. If one enters the two images, this case with that of pure fluid, the insertion of the porous layer causes thermal stability inside the reactor. There is no great difference in temperature between the active wall and the field, and the temperature varies generally in the range [453, 440 °C].

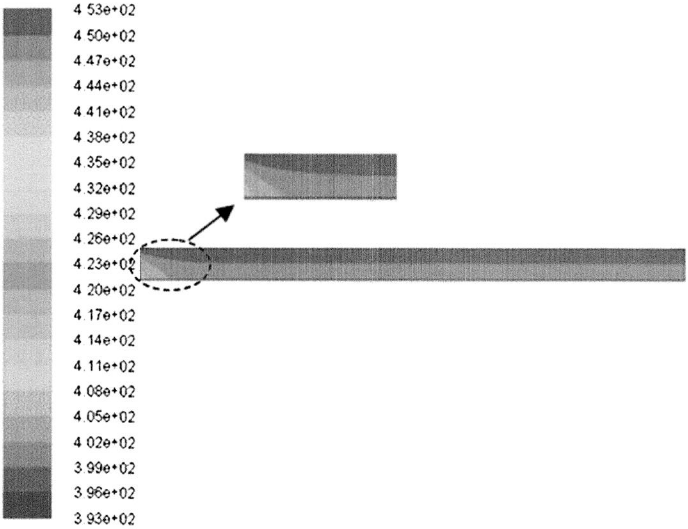

Fig. 4 Heat filed with metal foam inside the reactor

Chemical Species

Figure 5 shows the evolution of the molar fractions of the four chemical species: CH_3OH, H_2O, H_2, and CO_2. This image is qualitatively similar to that corresponding to the smooth channel. However, they differ in the addition of quantities of reagents and products in the porous case. The additional quantity is a result of the flow resistant caused by the presence of metal foam, and the matrix layer increases the residence time of reactant species within the reactor.

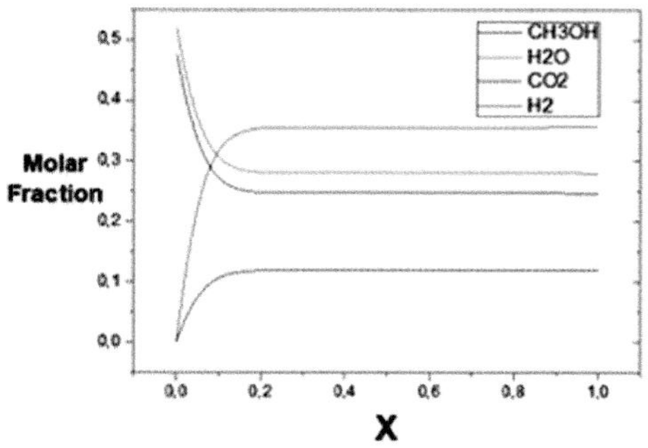

Fig. 5 Molar fractions evolution for all the chemicals species along the reactor

Reactor Performance with Matrix Layer

From the previous figures, the insertion of the porous media allowed the reactor to consume more reactants and consequently to generate more products, compared to the smooth channel reaction case. This is due to the good dispersion of chemical species and thermal energy, despite the resistance to flow (flow braking) which is generated inside the reactor.

The comparison between the two cases shows a stabilization of the temperature profiles in the second case (the temperature interval varied between 440 °C and 453 °C), this is due to the influence of the specific heat capacity of porous layer Cp_{MF} on the specific heat capacity Cp_{mix} (fluid–solid) combination, thus more consumption of gas mixture and more products. So, in this case, the yield becomes

$$\eta = \frac{M_{CH_3OH,in} - M_{CH_3OH,out}}{M_{CH_3OH,in}} = 49.5\% \quad (21)$$

Conclusion

This paper presents a numerical study by Fluent that treats the influence of a metal foam insert inside a methanol steam reformer, particularly on heat and mass transfer phenomena. The numerical results indicate that the metal foam is a good solution to lessen the rise of the fluid temperature inside the reactor, which cause the rise of the zone of the chemical reaction. The consumption of methanol and production of hydrogen are also increased by the insertion of the metallic foam. Therefore, the efficiency of the reformer is improved by 16%.

References

1. Pajak M, Mozdzierz M, Chalusiak M, Kimijima S, Szmyd JS, Brus G (2018) A numerical analysis of heat and mass transfer processes in a macro-patterned methane/steam reforming reactor. Int J Hydrog Energy 43(45):20474–20487
2. Wei Z, Yuzhi K, Qinghui W, Shaolong W, Jingdong L, Junpeng Z, Hui KS (2017) Development of cylindrical laminated methanol steam reforming microreactor with cascading metal foams as catalyst support. Fuel 191:46–53
3. Sanz O, Velasco I, Reyero I, Legorburu I, Arzamendi G, Gandía LM, Montes M (2016) Effect of the thermal conductivity of metallic monoliths on methanol steam reforming. Catal Today 273(C):131–139
4. Herdem MS, Mundhwa M, Farhad S, Hamdullah F (2019) Catalyst layer design and arrangement to improve the performance of a microchannel methanol steam reformer. Energy Convers Manag 180:149–161
5. Sanz O, Velasco I, Perez-Miqueo I, ´Poyato R, Odriozola JA, Montes M (2016) Intensification of hydrogen production by methanol steam reforming. Int J Hydrog Energy 41(10):1–10

6. Sari A, Sabziani J (2017) Modeling and 3D-simulation of hydrogen production via methanol steam reforming in copper-coated channels of a mini reformer. J Power Sources 352:64–76
7. Settar A, Nebbali R, Madani B, Abboudi S (2015) Numerical investigation on the wall-coated steam methane reformer improvement: effects of catalyst layer patterns and metal foam insertion. Int J Hydrog Energy 40:8966–8979
8. Morrison FA (2004) Compressible fluids. Michigan Technological University, Michigan
9. Wijaya WY, Kawasaki S, Watanabe H, Okazaki K (2012) Damköhler number as a descriptive parameter in methanol steam reforming and its integration with absorption heat pump system. Appl Energy 94:141–147
10. Santacesaria E, Carra S (1983) Kinetics on catalytic steam reforming of methanol in a CSTR reactor. Appl Catal 5:345–358
11. Jiang CJ, Trimm DL, Wainwright MS (1993) Kinetic study of steam reforming of methanol over copper-based catalysts. Appl Catal A 93:245–255
12. Agrell J, Hasselbo K, Jansson K, Jaras SG, Boutonnet M (2001) Production of hydrogen from methanol over Cu/ZnO catalysts promoted by ZrO2 and Al2O3. Appl Catal A 211:239–250
13. Horny C, Renken A, Kiwi-Minsker L (2007) Compact string reactor for autothermal hydrogen production. Catal Today 120:45–53
14. Du Plessis P, Montillet A, Comiti J, Legrand J (1994) Pressure drop prediction for flow through high porosity metallic foams. Chem Eng Sci 49:3545–3553
15. Boomsma K, Poulikakos D, Zwick F (2003) Metal foams as compact high performance heat exchangers. Mech Mater 35:1161–1176
16. Bai M, Chung JN (2011) Analytical and numerical prediction of heat transfer and pressure drop in open-cell metal foams. Int J Therm Sci 50:869–880
17. Poling BE, Prausnitz JM, O'Connell JP (2001) The properties of gases and liquids. McGraw-Hill, New York
18. Kaviany M (1991) Principles of eat transfert in porous media. Springer, New York
19. Nield DA, Bejan A (2000) Convection in pouros media. Springer, New York
20. Bhattacharya A, Calmidi VV, Mahajan RL (2002) Thermophysical properties of high porosity metal foams. Int J Heat Mass Transf 45:1017–1031
21. Chein R, Chen YC, Chung JN (2013) Numerical study of methanol–steam reforming and methanol–air catalytic combustion in annulus reactors for hydrogen production. Appl Energy 102:1022–1034

Open Porous Metal Fiber Structures for the Next Generation of Sorption-Driven Heat Pumps and Cooling Machines

T. Seidel, M. Fink, T. Hutsch and O. Andersen

Abstract Climate change demands the development of increasingly efficient heating and cooling technologies. Salt hydrates offer interesting hydration and dehydration temperatures and energies useful for energy-efficient heat pumps, sorption cooling devices, and thermal energy storage. As past research on zeolite heat pumps has shown, the combination of active material with advanced heat exchangers based on metal fibers leads to superior power densities for heating and cooling applications. A current project, therefore, investigates the combination of such heat exchanger technology in combination with different promising salts, like $MgSO_4$ or K_2CO_3, for application in sorption-driven devices.

Keywords Adsorption heat pump · Adsorption cooling · Metal salt · Fiber structure

Introduction

The supply of domestic and process heat in 2017 accounts for approximately 54% [1] of the German end energy consumption, which represents a large potential to reduce the primary energy demand and corresponding CO_2 emissions. Efficient heat transformation processes, e.g., by heat pumps, make use of ambient heat and low-temperature waste heat. By the application of reversible chemical reactions and physical sorption processes the necessary amount of electric energy can be reduced and, thus, the fossil energy demand caused by the electric energy mix. Additionally, these processes can be applied for cooling purposes, again reducing the fossil energy demand. A sketch of a continuous heat pump process is shown in Fig. 1.

Currently, the application of these sorption processes is not common, due to the challenges to accomplish an adequate heat and mass transfer which is, i.e., hindered by material agglomerations. Typically, porous media like silica gel are applied which

T. Seidel (✉) · M. Fink · T. Hutsch · O. Andersen
Fraunhofer Institute for Manufacturing Technology and Advanced Materials IFAM, Branch Lab Dresden, Winterbergstraße 28, 01277 Dresden, Germany
e-mail: torsten.seidel@ifam-dd.fraunhofer.de

N. Dukhan (ed.), *Proceedings of the 11th International Conference on Porous Metals and Metallic Foams (MetFoam 2019)*, The Minerals, Metals & Materials Series, https://doi.org/10.1007/978-3-030-42798-6_19

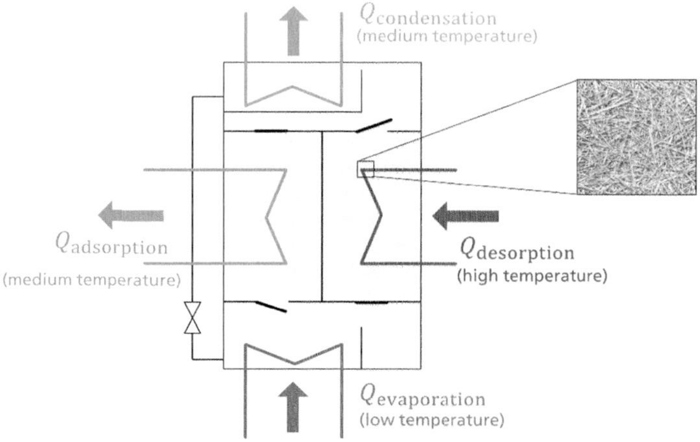

Fig. 1 Working principle of an adsorption heat pump with the application of a metal fiber adsorber

have a low thermal conductivity, thereby limiting the power density of these systems. The combination of a porous fiber structure, coated with a thin layer of metal salt adsorption material, provides an innovative solution to overcome these challenges and offers the potential for a broader application of the adsorption for heat pumps and cooling machines. The application of the porous metal fiber structures was already investigated during previous work [2] and showed promising results. An overview of the application of porous metal fiber structures for heat transfer applications is given in [3]. More details about the thermal conductivity properties are stated in [4].

Material Characterization on a Micro-Scale

An initial development step is the selection of an appropriate fiber material which offers the potential for a high thermal conductivity, optimum adhesion of the metal salt layer, a high material availability, and most likely a limited cost of the system. Depending on the applied metal salt, some candidates are shown in Fig. 2. The corresponding metal salts that can be produced are, for example, $Al_2(SO_4)_3$, $CuSO_4$,

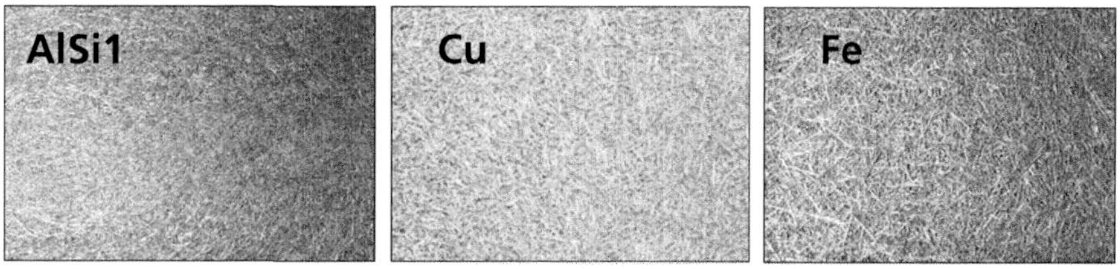

Fig. 2 Fiber structures made from potential adsorber materials

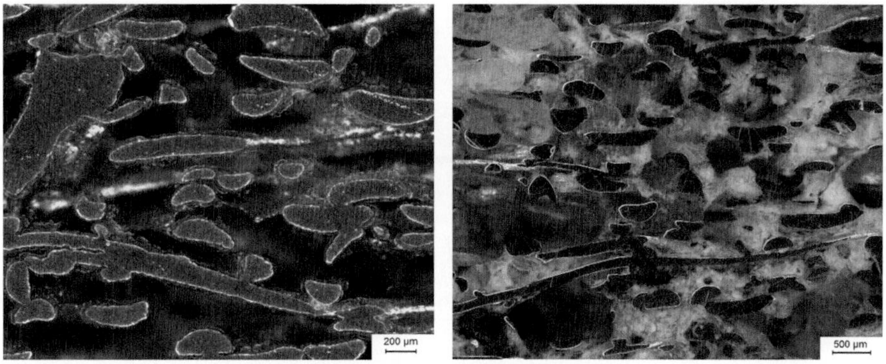

Fig. 3 Microsection of an AlSi1 fiber structure coated with $Al_2(SO_4)_3$. Left: thin layer. Right: thick layer

$CuCl_2$, and $FeSO_4$. Other metal salts were also investigated, but a sufficient fixture and compatibility with the fiber material are less easily achieved. The metal salt layers on the fiber structures were produced by the project partner Technische Universität Dresden.

The main quality factors of the coating are to provide an optimum fixture of the metal salt, a thickness that enables the application of a large amount of metal salt, distributed uniformly on the fibers, while still leaving enough space for the adsorbate (e.g., water vapor) to flow through the structure. Metallography is applied to assess the coating quality and thickness. Examples are shown in Fig. 3, which show the variations coming along with different coating technologies. Currently, the optimization of the coating process is still in progress.

The composition of the produced metal salt layers was analyzed by X-ray diffraction (XRD) to identify the chemical substances that were produced at the surface during the coating process (reaction between metal and a suitable acid) and ensure that the right metal salt hydrate was prepared. An example for $Al_2(SO_4)_3$ is shown in Fig. 4. Different species of aluminum salt hydrates have been identified and indicate that an adequate sorption material has been produced.

Material Characterization on a Macro-Scale

The previously described analyses have been supported by experiments to determine the thermal and flow properties as well as thermogravimetric experiments to determine reaction equilibria (temperature and pressure). The results are applied as a basis to design a lab-scale demonstrator to assess the technology under practical conditions. The heat and mass transfer in the fiber structure is influenced by its heat conductivity and the pressure drop along the flow path through the coated structure. Pressure drop and heat conductivity are, in turn, influenced by the fiber properties

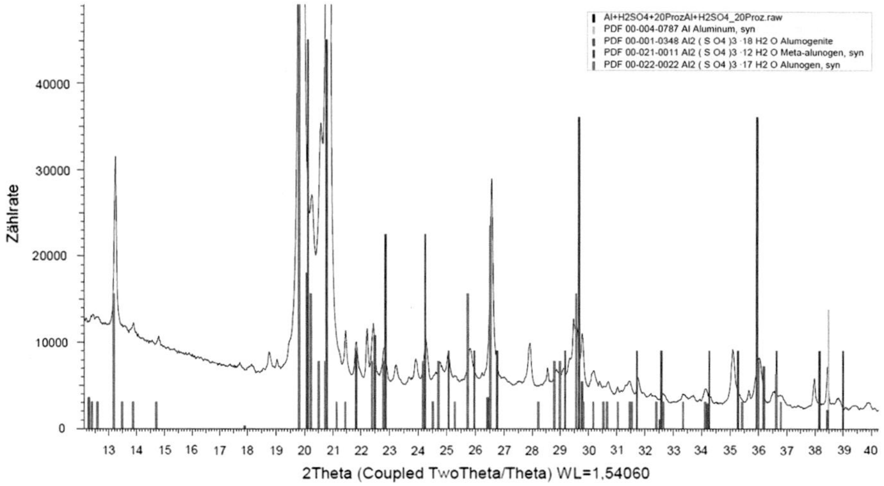

Fig. 4 XRD of an AlSi1 fiber structure coated with $Al_2(SO_4)_3$

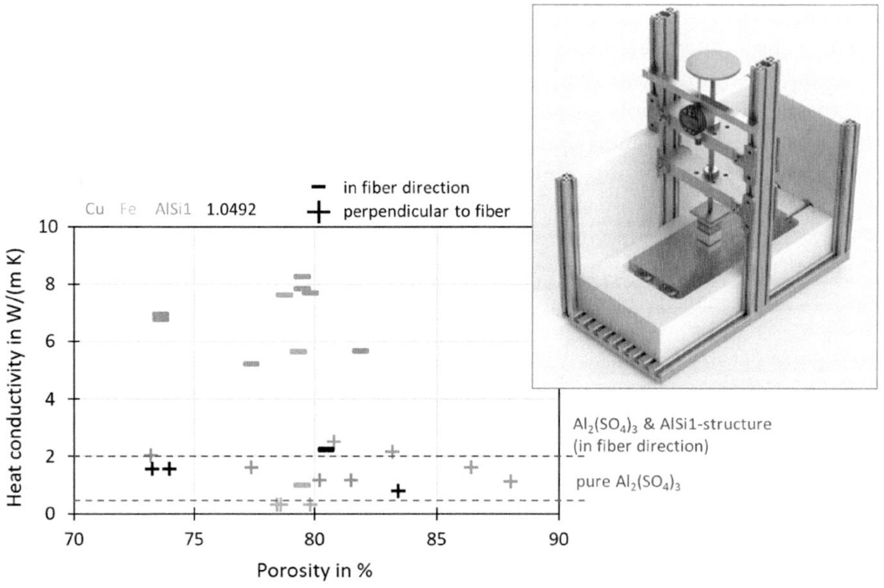

Fig. 5 Experimental setup for the measurement of the heat conductivity and measurement results for different fiber structure porosities and materials

(diameter, specific surface area), the total porosity of the fiber structure, and the coating thickness. Only a suitable combination of all of these properties will result in an optimum overall performance.

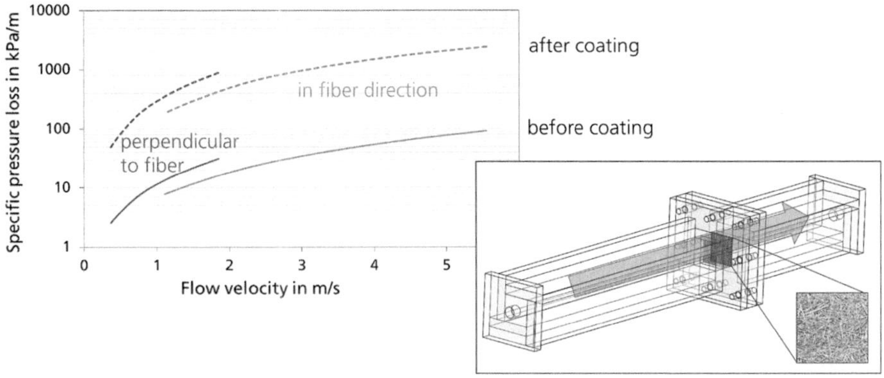

Fig. 6 Experimental setup (flow channel) to determine the pressure drop and results for an uncoated and coated AlSi1 fiber structure

The thermal conductivity was measured with a plate apparatus, shown in Fig. 5, according to ASTM D 5470-2017. Sample size was approximately 30 × 30 × 10 mm^3. The results for different fiber structures are shown in the same figure. The reference case is the thermal conductivity of the pure metal salt, represented by the red dotted line for the $Al_2(SO_4)_3$. Depending on the numerical value of the porosity, almost all uncoated fiber structures show a higher thermal conductivity than the pure metal salt. The overall thermal conductivity of a metal salt $Al_2(SO_4)_3$ coated AlSi1 fiber structure (purple dotted line) has a lower value than the uncoated fiber structure; however, the averaged thermal conductivity of the coated sample is more than four times higher than that of the pure metal salt. This verifies the improvement of the thermal conductivity by the fiber structure.

Regarding the mass transfer in the structure, the pressure drop of the structures has been determined with a custom-made flow channel, shown in Fig. 6, using an ALMEMO® data acquisition system and pressure drop sensors. The same samples as for the heat conductivity measurements have been tested. The most important result of the experiments is the pressure drop increase due to the coating, shown also in Fig. 6. For the same sample, as described in the heat conductivity measurements, the pressure drop increases by a factor of approximately 27 due to the coating. Thus, further optimization of the coating with regard to achieving homogeneous thin layers without blockage of the flow paths is desirable.

The thermogravimetric analysis has been performed with a Netzsch STA 449. The aim of the experiment was to obtain experimental data for the temperature and pressure conditions for the dehydration to supply the data to the partner Technische Universität Dresden, who collected corresponding literature data for comparison of the results. The sample mass was typically in the range up to 50 mg. For the experiment, a steam generator was applied to regulate the humidity. The carrier gas was nitrogen to prevent chemical reactions caused by the gas. An exemplary experiment for $MgSO_4$ is shown in Fig. 7. Different dehydration steps can be identified and

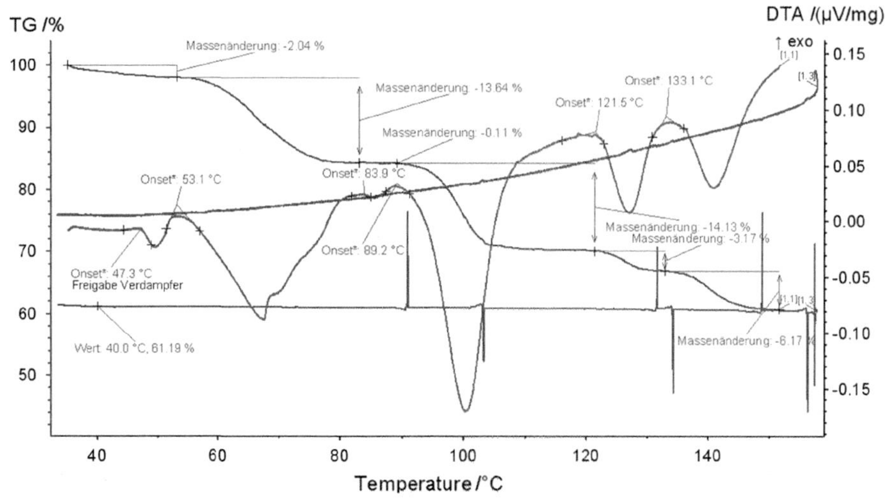

Fig. 7 Thermogravimetric analysis of an $MgSO_4$ hydrate during desorption

the corresponding change in mass allows to determine the amount of water being desorbed. Based on this data, a demonstration adsorber can be designed.

Conclusions

The micro- and macro-scale analysis provided the basis for the design of a prototypic adsorber system which requires a stable and reliable coating. The principle of the adsorber setup is shown in Fig. 8 where the fiber structures are stacked on a tube which is mechanically widened in order to provide good heat conduction between the tube and the fiber structures. This represents a scalable and established concept to design an adsorber and was already demonstrated for latent heat thermal energy storage prototypes. The heat transfer and pressure drop measurements allow the identification of optimum parameters for the porous fiber structure.

The next steps include the selection of an appropriate metal salt for the demonstrator, adaptation of the coating technology to the larger part geometry, identification of optimum fiber structure parameters, and building of a system for practical testing under application-relevant conditions for several adsorption and desorption cycles to demonstrate the ability to deliver a high volume-specific heat power output. Based on a successful demonstration, the system can be scaled up and transferred to an industrial scale based on future product-related developments.

Fig. 8 Experimental setup (flow channel) to determine the pressure drop and results for an uncoated and coated AlSi1 fiber structure

Acknowledgements Funding from the Federal Ministry for Economic Affairs and Energy for the project 03ET1451C is greatly acknowledged.

The coatings with the metal salt were prepared by the project partner Technische Universität Dresden, Germany and are also greatly acknowledged.

References

1. German Federal Ministry for Economic Affairs and Energy, "Energiedaten: Gesamtausgabe", August 2018
2. Wittstadt U, Füldner G, Andersen O, Herrmann R, Schmidt F (2015) A new adsorbent composite material based on metal fiber technology and its application in adsorption heat exchangers. Energies 8:8431–8446
3. Fink M, Andersen O, Seidel T, Schlott A (2018) Strongly orthotropic open cell porous metal structures for heat transfer applications. Metals 8:554
4. Veyhl C, Fiedler T, Andersen O, Meinert J, Bernthaler T, Belova IV, Murch GE (2012) On the thermal conductivity of sintered metallic fibre structures. Int J Heat Mass Transf 55:2440–2448

Author Index

N. Dukhan (ed.), *Proceedings of the 11th International Conference on Porous Metals and Metallic Foams (MetFoam 2019)*, The Minerals, Metals & Materials Series, https://doi.org/10.1007/978-3-030-42798-6

Subject Index

N. Dukhan (ed.), *Proceedings of the 11th International Conference on Porous Metals and Metallic Foams (MetFoam 2019)*, The Minerals, Metals & Materials Series, https://doi.org/10.1007/978-3-030-42798-6

V

Z

GPSR Compliance

The European Union's (EU) General Product Safety Regulation (GPSR) is a set of rules that requires consumer products to be safe and our obligations to ensure this.

If you have any concerns about our products, you can contact us on ProductSafety@springernature.com

In case Publisher is established outside the EU, the EU authorized representative is:

Springer Nature Customer Service Center GmbH
Europaplatz 3
69115 Heidelberg, Germany

Batch number: 10756376

Printed by Printforce, the Netherlands